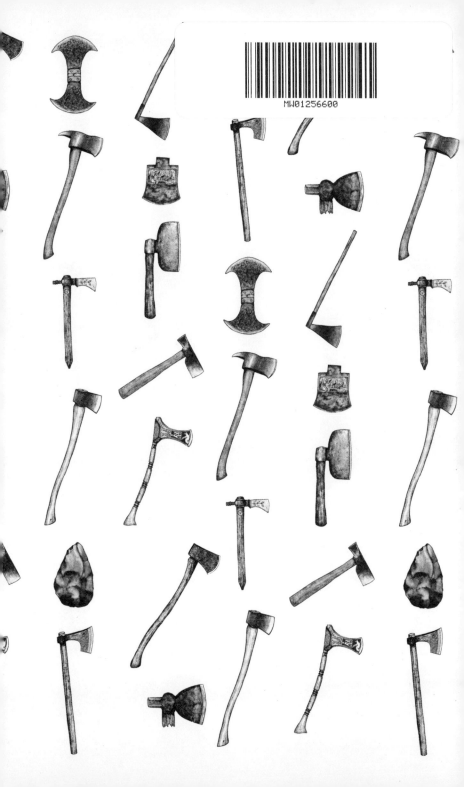

WHACK JOB

WHACK JOB

A HISTORY OF AXE MURDER

RACHEL McCARTHY JAMES

ST. MARTIN'S PRESS
NEW YORK

First published in the United States by St. Martin's Press,
an imprint of St. Martin's Publishing Group

WHACK JOB. Copyright © 2025 by Rachel McCarthy James.
All rights reserved. Printed in the United States of America.
For information, address St. Martin's Publishing Group,
120 Broadway, New York, NY 10271.

www.stmartins.com

Designed by Steven Seighman

Endpaper illustations by Rhys Davies (rhysspieces.com)

The Library of Congress Cataloging-in-Publication Data is available
upon request.

ISBN 978-1-250-27673-5 (hardcover)
ISBN 978-1-250-27674-2 (ebook)

Our books may be purchased in bulk for promotional, educational,
or business use. Please contact your local bookseller or the
Macmillan Corporate and Premium Sales Department at 1-800-221-7945,
extension 5442, or by email at MacmillanSpecialMarkets@macmillan.com.

First Edition: 2025

10 9 8 7 6 5 4 3 2 1

Dedicated to the authors of endnotes 1 and 2, chapter 9

CONTENTS

INTRODUCTION: Nothing Could Be Simpler..........................1

Stone Hand Axe..18
ONE: Cranium 17 and the Pit of Bones..............................20

Type G-VII Egyptian Battleaxe... 32
TWO: The Smited King 35

Yue 47
THREE: Cascade of Blood 50

Pélekys... 57
FOUR: In Truth, an Enemy and a Man of Violence................59

Iron Shipbuilding Axe... 71
FIVE: Freydis, Woman of the Forest 73

Executioner's Axe...88
SIX: Pigmen, Gargoyles, Blundering Youths.............................90

Tomahawk..112
SEVEN: You Are Not Dead Yet, My Father.............................114

Cooper's Side Axe..129
EIGHT: I Suppose You Know What I Am Doing...................131

Roofing Hatchet...139
NINE: Five Axes in the Cellar, One Axe on the Roof............142

Shingling Hatchet..162
TEN: A Fuller Measure of Life and Truth, at Any Cost..........164

Boy's Axe..180
ELEVEN: "Whoever Comes Over, I Give Anybody Candy"......182

Felling Axe..195
TWELVE: Candy...197

Fire Axe...216
EPILOGUE: Crime and Difficult Situations..........................221

Acknowledgments...233
Selected Additional Sources 239
Notes 247

As an axe leaves its mark in the speechless tree, so an evil deed leaves its mark in the evil doer's consciousness.

—John Henry Wigmore, as quoted in Cara Robertson's *The Trial of Lizzie Borden*

My axe is my buddy, we both cry with the trees
Me and my axe will bring the devil to his knees

—Insane Clown Posse, "My Axe"

WHACK JOB

INTRODUCTION

Nothing Could Be Simpler

Early on in the process of writing my first book about axe murder, I took a job working in an after-school program. I was twenty-six and I'd recently figured out a new answer to the hundred-year-old mystery of the murder of eight people in Villisca, Iowa. I needed a day job, and the children at the nearby elementary school I had once attended myself were a bright counterpoint to the grim work of researching true crime.

Mornings spent mired in awful murders were balanced by cheerful afternoons in sunny playgrounds and colorful classrooms, mediating disputes over nothing more serious than iPad time. At Pinckney Elementary, my job working in after-school care with the Boys & Girls Club was both more and less important than my work unearthing major connections between horrific tragedies. These kindergarteners had real struggles too, but I wasn't a teacher or their parents.

Anything major was above my pay grade. I had to make sure that the kids didn't die, but only in the two or three hours they were in my care.

One day a little boy said something that struck me. The setting was the gym, a routine safety assembly. Fire was banal as a threat, routine, done to death. But in 2013, the possibility of a mass shooter was still novel and exciting enough to make the kids antsy. Especially since we were talking about school shootings like they were some forbidden R-rated movie. Barricading doors, running away, hiding, all of those things were covered—but we avoided mentioning the actual threat, the weapon, the gun.

The boy, a fifth grader, raised his hand, blurting out, "WHAT IF IT WAS AN AXE MURDERER?" He accompanied the question with a mimed motion and sound effect that was somewhere between light saber and baseball bat.

We just moved on without reacting—you can't rise to the bait that way when you're trying to get children to take something seriously. But the incident stuck with me. How did this child, born into the age of touch screens, have a reference point for "axe murderer"?

It makes sense that he wouldn't say "what about a SCHOOL SHOOTER?" since that was the source of the very tension that the joke intended to relieve. Instead, he reached into his imagination and came up with a tool. Axes are a big deal in *Minecraft*, so I have no doubt that he understood that their primary purpose as a tool is to cut down trees. Axe murder, though, seemed a joke too dark for such a tender age. How was the violence of an axe murderer remote enough to be funny and yet familiar enough for a ten-year-old to invoke?

The boy's antics stuck with me throughout the process of writing and promoting my book *The Man from the Train*. Like this book, it is a work of nonfiction. The man in question would ride the rails to a community where he had no ties, pick up an axe left outside in a woodpile, break into a house, and kill everyone inside; he especially targeted households with children. The book makes the argument that this pattern repeated dozens of times nationwide over about fifteen years and can be traced back to a man who evaded capture after murdering the family he lived with and worked for on a small farm in Massachusetts. From the beginning of my research process, the idea of the *axe murderer* as opposed to a murderer who used an axe was central to my understanding of these tragedies.

Newspapers embraced the grabby compound phrase slowly: in 1890, the term was almost never used, but by 1920 it was used frequently. Our fiend in *The Man from the Train* killed perhaps as many as a hundred people beginning in 1898. He began relatively slowly, slaying one or two families a year, choosing rural communities and leaving geographic distance between his attacks. We believe that he was following work opportunities—especially the chance to use the axe to fell trees as a lumberjack. In 1909, he ramped up the bloodshed and the events became more frequent and more tightly spaced. Though the idea of serial killers wouldn't be formalized for decades to come, awareness of them was dawning among the crime-obsessed public. After the man killed two

Colorado Springs families on a September night in 1911, headlines about the "axe-murder"[1] began to take root in newspaper pages, especially in the Midwest.

The coverage of that event died out, but the phrase took hold with the trial of Clementine Barnabet. She was a young black woman in Lafayette, Louisiana, a troubled eighteen-year-old obsessed with cults and true crime who confessed to a series of murders in Louisiana and Texas—horrible crimes she (in my firm opinion) did not commit. Her vivid court statement involving voodoo rituals grabbed headlines across the country; white journalists found it easy to exploit their readers' racism by placing exaggerated and obvious rumors into print. By the time the man from the train made his infamous trip to Villisca in June 1912, the phrase "axe murder" was a well-established part of the crime-writing lexicon, never to exit it again. It stuck to the Villisca event so firmly that when I first started research in 2012, the Wikipedia page for it was the first Google result for the phrase "axe murder."

I once believed that the emergence of this wording in newspaper headlines also mirrored a surge in the use of axes themselves as a weapon, but now, having tracked the axe over so long a period throughout human history, I don't think that's true. Axes are at their core utterly common, and so they have been as ubiquitous as weapons as they have been as tools. They are everywhere in our story of violence, so much a part of the texture of our conflicts that they become banal, unworthy of note.

What made the phrase "axe murder" so charged in the

1910s was a clash of cutting-edge technology with ancient traditions in toolmaking over the previous sixty years. Industrialization would eventually make the axe fall out of daily use, but for a moment the new machinery made the humble axe omnipresent.

Battleaxes were thoroughly outdated by the Civil War, but in military and domestic settings there were more axes than perhaps at any other point before in history. Steel tool use proliferated after 1856, when a man named Henry Bessemer devised a new method of purifying iron quickly and inexpensively by oxidizing iron with pressurized air to remove carbon and other impurities. This one-step method was discovered in the process of producing cannonballs that could be fired in the manner of a rifle. Bessemer was driven to problem-solve after the British postal service stole his fiancée Ann Allen's idea to modify the dates on embossed postal stamps, a grudge he held even after his steel foundry made him wealthy. Steel had previously been quite precious and rare, but unlike lab-grown diamonds its new availability and cheapness did not temper demand. Skyscrapers and trains and paper clips alike flowed out of the foundries, along with a flood of hand tools: hammers, shovels, hoes, and others were suddenly much more common and inexpensive. In the 1866 Fyodor Dostoyevsky masterpiece *Crime and Punishment*, the antihero Raskolnikov murders an elderly pawnbroker and her sister with an axe, grabbed from the kitchen in his apartment building. In the depths of his absurd and monstrous plans, in the chaos of Saint Petersburg and his own head, he settles upon the convenience of the axe: "nothing could have been simpler."[2]

As electricity and plumbing and cars and telephones became increasingly accessible, the axe seemed all the duller. Guns supplanted the axe as the convenient weapon of choice. Chainsaws were used for a larger and larger share of forestry applications, though they weren't yet in use in the home. Axes were becoming an object of slight kitsch. At the 1893 World's Fair, a small glass version of George Washington's axe was a popular souvenir. Carrie Nation—a leader in the temperance movement of the early twentieth century who raided and destroyed bars and speakeasies—remained an object of derision long after the repeal of prohibition, her tiny hatchet no equal for the country's thirst. One brand of tobacco in the form of a plug (a brick of compressed tobacco leaves, to be cut off and chewed or smoked in a pipe) called itself Battleaxe, recalling a now firmly extinct era of violence to flatter male customers who want to think of themselves as rugged—not unlike the logic behind Axe personal care products today. The axe was still around, but becoming antiquated.

By 1922, the phrase "axe murder" appeared in U.S. newspapers hundreds of times each year, always used to describe actual crimes. But, like all grabby phrases, it became tinged with the grime of exploitative newspaper coverage through overuse. As the chainsaw took more and more of the axe's main occupation, the axe became an anachronism.

The contrast between the axe (boring) and murder (exciting) gave the phrase a darkly comic tinge. By the 1950s, "axe murder" was still a staple of newspaper headlines, yet the phrase and indeed violence with axes itself was becoming a stock joke. In Shirley Jackson's lighthearted book of domestic essays *Life Among the Savages*, she writes of a sleepy domestic scene perked up by violence: "I finally found an account of an axe murder on page seventeen, and held my coffee cup up to my face to see if the steam might revive me."³ In 1955, Charles Schultz killed off unpopular character Charlotte Braun in an illustration that parodied the gentle humor of *Peanuts* by putting an axe in Charlotte's head. In the 1964 hagsploitation movie *Strait-Jacket*, Joan Crawford campily raised the axe for B-movie thrills to theatrical success.

Axe murder had primal connotations, deeply interpersonal and separate from the state violence of execution rituals and battlefields where axes have created the most bloodshed. These associations came in handy in 1976, when North Korea and the United States nearly got into a war about a tree. In the North Korean axe murder incident, Captain Arthur Bonifas and Lieutenant Mark Barrett of the U.S. Army tried to trim a poplar tree that had created a surveillance blind spot. The tree grew on the Bridge of No Return, a tense spot near the Demilitarized Zone where North Korea, South Korea, and the United States observed an uneasy truce, often taunting one another by banging on windows with axe handles. The captain and lieutenant were killed by thirty North Korean soldiers, armed not just with their own axes but with "crowbars, pipes, [and] clubs" taken from South Korean laborers.⁴ The United States responded with restraint by simply cutting down the tree (with a much more efficient chainsaw), leading to some crucial concessions from the North Korean president and transforming the tripartite relationship. Though the event was not especially well remembered, it illustrates some of the crucial associations of axes and violence in the United States. The term "axe murder" characterized the incident as a close-quarters grudge, not a military exercise with guns and helicopters but something ugly and outside of bounds. And, since this war was almost begun by trimming a tree and ended with chopping down that tree, there's a sense of something just barely funny about it.

Axe murder—or really, axe murderers—became a permanent punchline after *The Shining* in 1980. The moment where Jack Torrance hacks through a locked door with a fire axe is darkly funny all on its own, with its invocation of Johnny Carson's signature catchphrase; the shot itself harkens back to a movie called *The Phantom Carriage*, made around the first flush of "axe murder" in 1920. When a Mike Myers movie goofily satirizing the eighties erotic thriller needed a punchier title, they went to the *Shining* parody set piece at the center and called it *So I Married an Axe Murderer*. Though the comedy wasn't an initial hit, it's become an enduring cult classic, permanently sealing the idea of axe murder as comic. Infinite references, parodies, homages of the axe breaking through the wood in *The Simpsons* or *MythBusters* or Super Bowl ads for Mountain Dew Zero Sugar have built a long legacy for the fire axe wielded by Jack Nicholson. But the phrase mostly disappeared from newspapers describing actual crimes. It is now too unserious.

The axe in silhouette still has genuine menace when the tone is dark, and power—an axe, usually with a hyperbolically huge blade, is often a formidable weapon in violence-oriented video games. But an axe murder is also appropriate for a throwaway line in a lighthearted family comedy like *Bob's Burgers*, the second of three in a list: "roller coasters, or axe murderers, or Dad's morning breath."[5] An escalation from roller coasters on the way to the real punchline. The idea of an axe murderer is not so much a shock as a joke so easy a ten-year-old could make it.

Axes are far from humanity's only deadly, omnipresent instrument. Hammers and swords and knives are objects of fetishistic fear, each in many ways more iconic than the axe. Knives and hammers are an even more essential tool for the little jobs of daily life, closer at hand than an axe in most modern households. And while axe throwing has had a recent upswing in popularity, fencing is an Olympic sport. Axes can be expensive toys and objects of honor and prestige, but a hatchet will never have the pedigree, the classiness, the pretensions of the épée.

Yet it is that earthbound flintiness that makes the axe so iconic. There is no tool older. The axe is the sword, it is the hammer, the planer, the scythe, khopesh, knife.

As every task somehow involves your smartphone today, so everything required a hand axe when humanity began. Early people spent their lives refining their craftsmanship to make these biface tools beginning about two and a half or three million years ago, learning the qualities of different kinds of stones and perfecting the flaking and percussive techniques to make these elaborately faceted gems into teardrop shape. And then half a million years ago, the tools got a hell of a lot better. Handles were a miracle of engineering, allowing the one-piece blade to do "cutting, scraping, chopping, and piercing."[6] This was a transformation in hunting and in interpersonal violence.

The axe gained literal power with the handle, its force physically enhanced by the new leverage, and became all the more valuable as a tool, as a weapon, as a symbol of wealth. Specialized axes spread into professions beyond woodcutting, such as coopering (barrel making), shipbuilding, and roofing, and revealed their usefulness for enforcing community disputes.

The axe's power only grew with the introduction of metals. Otzi—the naturally mummified man who lived 5,300 years ago in the Alps of Northern Italy and was discovered in 1991—held an axe. The head of the tool was fitted into the handle and lashed rather than the other way around—typical for stone axes. But this blade was copper, a sign of the man's high social status, and his weapons were flecked with the blood of others.

By the time we start to have recorded history, the axe is already loaded with the language of power. The Dresden Codex—a Mayan book that is one of the oldest in the Americas—had a specific ideographic symbol, *ch'ak*, defined as the "axe/comb glyph,"[7] which possibly refers both to events of war and decapitation as well as the movement of the planets. This association is reflected in the warrior god of lightning Chahk's great jade axe—an association between the axe and the great whacks from the sky that persist across many systems of belief from around the world: Pangu's axe that cleaved the world in Chinese mythology, or Hephaestus hitting Zeus's head with an axe to birth Athena.

State violence invested the axe not just with power but with authority. In war, execution, and other official and sanctioned shows of power and force—the axe was always present in the chaos of battle and the awful order of ritual violence. Battleaxes were thinner than woodcutting axes, but both were present on campaigns, key to battles on land and sea. Axes were a distinctive part of many legendary military forces, even into the twentieth century in some cases—in the Democratic Republic of the Congo, the Songye people made elegant hatchets for ceremonial decapitation rituals that have spokes where the cheek of the blade would usually be, etched with human faces.

The posture a person must assume to be beheaded is submissive. To lay your neck on the block means that you have not just been defeated but that you have stopped fighting. This is the theater of death: it is not so much about making the condemned feel their defeat, but about communicating the totality of their defeat to the people watching, who might wish to succeed where the condemned failed. This is the reason the axe was such an important part of the fasces, a symbol from Rome that has resonated with genocidal imperial empires from the United States to Germany, and the namesake of fascism. The axe tells anyone who hopes to resist that they will be not only killed but also ideologically defeated.

But even as the axe can be put toward the abuse of power, so it can be put toward resistance. The commonness of the axe and its inseparability from the rank and file of labor processes lend it unexpected power in rebellion and revolution. The artist Kerry James Marshall has returned again and again to the subject of Nat Turner, who led a slave insurrection armed with axes in southern Virginia in 1831; the axe appears in two of these paintings, for woodcutting and for beheading Turner's enslaver.*

In the establishment and westward expansion of the United States, the axe became pivotal. Early on, it was heavily taxed as an import, making it a scarce and valuable tool in trade. Demand only grew. This was once a heavily forested country full of unimaginably huge trees, and there were homes and farms and railroads to build. Even nonrural families like the Bordens found themselves with five different axes in their cellar. As the Bessemer method led to the explosion of steel tools in general, there were suddenly many more patterns of axes to fit different needs: roofing, woodcutting, firefighting. The Pulaski, an axe with a hoe on the back of the pole invented in 1911, endures especially as an essential firefighting tool.

---

* Marshall also painted a version of Julian Carlton, the axe murderer of our tenth chapter. Curiously, *The Actor Hezekiah Washington as Julian Carlton Taliesen Murderer of Frank Lloyd Wright Family* depicts a man with no visible hatchet. Perhaps this relates to Carlton's silence in the days between the murders and his death, while Turner's confession reverberated loudly.

But by 1965 the axe was no longer the default choice for so many of the tasks it once handled. Chainsaws took its truest purpose, woodcutting. Electric heat reduced the need for firewood cut in the home. Grocery stores made it easier and easier to buy meat that was already butchered. Cars meant there were no more carriages to repair or horses to shoe. Today, you only have five axes in the basement if you *really* like axes.

It's not that the axe is no longer useful. Someone who is sleeping rough in the woods has great need for a hatchet to clear a place to sleep. A housed person with a fireplace or a backyard full of scrubby trees could use an axe toward those purposes; there are less work-intensive options, but the challenge and the connection to the old ways are the point. Like the cowboy or the gangster, the woodsman has turned from a laborer into a mythical beast of masculinities past, a Paul Bunyan fantasy you can buy at the hardware store. And that's not silly; it fulfills something deep within us. But the axe doesn't carry the connotation of practicality that it once did—it's usually not the most efficient way to solve whatever problem is in front of you.

Especially violence. Guns are easy to use, there are a lot of them, and they're much likelier to be used to fatal effect than the axe. And yet the axe is still a common object that gets picked up in unpremeditated violence, whether in the great outdoors or the garage. Though gun purchase laws are nowhere near tough enough, it's still much easier to buy an axe than a gun.

In this book I've tried more than anything to look directly

at the axe, both as it gets lost in a sea of other tools and technology and as it becomes a fetish or an afterthought in violent stories throughout time. I've reviewed true crime staples, toured three Frank Lloyd Wright buildings, hiked a Pennsylvania glen, and dug into the backstory of a throwaway sentence in Herodotus. The kind librarians at the University of Kansas sent scores of books about everything from handles to sex crimes to Vikings, right to my office at my day job (and let me renew them for years). Lost one summer on a college campus in Memphis, Tennessee, I found a huge mural of Ramses II smiting his people with a circular axe. And I've learned to see human-sacrifice rituals everywhere.

Each story in this book is prefaced with a brief explanation of one specific kind of axe, and how each new blade reflects the axe's evolution as a piece of technology, weapon, and cultural symbol. There are hundreds and sometimes thousands of years between many of these murders. The events in these books take place at such disparate moments in history, but they have commonalities beyond the tool central to each: revenge, guilt, entitlement, exploitation, shame, spite, war, greed, madness.

And occasionally, freedom, resistance, redemption. Community, even.

Once I began to study the blade, I found that the object of my attention unlocked a new level in the world around me. When I watched television, any appearance of an axe was cause to exclaim—doubly so if it was an actual reference to axe murder. When I saw an axe-throwing place in a little middle-of-nowhere town, obviously it was a sign. When I went to art museums, I made a scavenger hunt of finding all the axes I could, in the weaponry room or in the background of bucolic landscapes. My friends started sending me axe stuff, and I've started collecting axes and related tchotchkes, the way my mom collects little elephant statues. It feels as if every axe is a reference to me. The reason it felt like a secret new language, the reason I felt like no one cared about it but me, is because most people take it so much for granted. We take for granted that which is everywhere, inextricable from the landscape of the forest, of the utility shed, of violence.

Like anyone who declares a number one hit to be their song, the axe is not actually my thing. It's a popular hobby, and a lot of people know a lot more about how to use it than me. Collecting, using, and maintaining axes can be an expensive hobby, and a lot of the people who are very into it are into it for better reasons than me. I love the axe as a symbol, but they understand it on a deeper level, not just as an intellectual pursuit but as a physical and practical application. Even after several years working on this project, there are a ton of details that I can't retain but that seem to be second nature to the redditors on r/axecraft. I care about the difference between a tomahawk and a Hudson Bay hatchet, but the difference between a Baltimore Kentucky and a Delaware Jersey pattern axe-head is not so important to me. Axes are a fun hobby because there are thousands of minute yet important variations, but also because they're just fun. Fun for collecting, fun for exercise, fun for sport.

Axe throwing seems dangerous, but not to the point where you won't have a beer and joke around while you're doing it. A sharper thrill than bowling.

No one has died in an axe-throwing accident yet, though it seems inevitable. One time in a hotel room I came across an axe-throwing competition on cable in which all the advertisers were insurers, holding their breath. The axe's simplicity is armor against the worst-case scenarios. No trigger, no bullet, no rotating set of teeth, just *One Moving Part*, as the U.S. Forest Service titles its loving and detailed manual of the axe.

And yet the axe is so sharp and it needs such maintenance that it becomes a tool of accidents, missing toes, lost fingers. That's the scenario in a verse in Deuteronomy that has fascinated me since I found it near the beginning of this project. Verse 19:5 reads:

> *For instance, a man goes with his neighbor into a grove to cut wood; as his hand swings the ax to cut down a tree, the ax head flies off the handle and strikes the other so that he dies. That man shall flee to one of these cities and live.*[8]

The guidance is surprisingly nuanced for a stark chapter in the Torah that ends: "[Show no] pity: life for life, eye for eye, tooth for tooth, hand for hand, foot for foot."[9] It is presupposed with a clarification that the "manslayer" was not already an enemy, a complication for a hypothetical explanation of newly set laws. Sort of a fifth century BCE version of a question on the bar exam. Here the axe is a tool of the workingman, not the murderous fiend, and yet still a slayer.

Blood feuds were one of the main ways to deal with the fallout from a violent death in the culture that produced the Torah. Cities of refuge were a way to levy a cost for the grief of an accidental but terrible and unjust death. The man whose hands held the tool that killed another still must answer for the death they didn't mean to cause. Not with their own life, but with the life they've built in their community. That life, the one that included their poor neighbor, is over—but they can rebuild.

The axe is a tool of construction and progress, an instrument of growth easily turned into a method of death. In seeking to understand the tool, I seek to understand why it was picked up by the killers who drew blood by its blade and poll, who struck fear just by hoisting it aloft. I do not have refuge to offer the axe murderers or the survivors or the victims, just a connection between these deaths and our lives. Hold the axe for a moment, press a cheek to the cold cheek of the steel head, find a grip on the handle. Press my thumb to the sharp blade and feel the cutting edge do its work.

Stone Hand Axe

430,000 BCE, Northern Spain

The first axe looked very different from the axe on the cover of this book. Early axes were made from rocks, usually flint, diligently chipped away at to form a large tool. Sometimes the tool was symmetrically shaped, but sometimes one side was flat and the other was concave, making it easier to wield in both hands. The body narrowed from a wide base, sometimes thinning to a tip sharp enough to pierce and sometimes just barely funneling to a rounder top, forming a slightly unbalanced oval. The blades on the side were the important part, the strongest link to the axes you'd find at the hardware store. They were serrated rather than smooth, but sharp enough to scrape, slice, or stab.

And they were everywhere, transmitted from culture to culture like a whetted disease, "one of the commonest, most widely distributed, and longest-lasting archeological artifacts."[1] The earliest stone tool dates back as far as three million years, from eastern Africa. These Acheulean-era triangular, tear-shaped bifaces with sharp edges, flaked on both sides, began to appear in Kenya about 1.6 million years ago. Hand axes spread north, arriving in Europe by 900,000 BCE. They are only missing when there was no human activity in the area; one influential theory of early human history, called the Movius Line, concerns the lack of these hand axes in eastern Asia compared to Africa, Europe, and southern and western Asia, due to the climate and insurmountable natural obstacles like the Ganges. Wherever people went, they showed each other how to make rocks into axes.

So many archaeological insights are gained through practice. We learn how these axes became sharp by trying to re-create the conditions of their sharpening: hitting and throwing these rocks against bigger rocks. And we understand how they were used by using them again: to cut animal skins, level out wood surfaces, crush and break seeds and nuts and grains. But our research has its bounds: we're not going to use these newly knapped axes to hit someone's skull and see exactly how it cracks.

ONE

CRANIUM 17 AND THE PIT OF BONES

Murders do not always announce themselves. Accidents can be as bloody and upsetting as any homicide, and are much more common. Misadventure and malice are easy to confuse.

This is particularly true with the passage of time. The farther the corpse travels, the drier their blood, the more decayed the flesh. In short, the deader a person's body gets, the harder it is to demystify what happened to make them dead. Long before they are reduced to a skeleton, the person they were lost their ability to narrate their own death, revealing whether the culprit was a stranger with an axe, a husband with a baseball bat, or a simple fall down the stairs; there are so many ways to die suddenly. But the vacated skulls still have stories to tell.

The hat-brim-line rule is the name for a rather rough measure of whether an impact to the skull came from an accidental injury or lethal interpersonal violence. Most hats cover the base of the occipital to the forehead, just above the brow line: imagine a line separating the parts of your head covered by a baseball cap and the parts that are not. The bare parts are the parts of the head that aren't usually going to get hurt unless someone wants to hurt them. It's the prettier parts of the head that are most vulnerable in an accident. Falls, crashes, collisions—most of these things happen when you're facing forward.

When there's body fluids and murder weapons and circumstantial evidence to consider, the hat-brim-line rule isn't of much use. It's way too unreliable to soundly convict a person: drunkenness or brain injuries or the vulnerabilities of extreme youth and old age can explain many impacts to the back of the head, and when the injury is right on that three-inch hat-brim line, it's hard to differentiate one way or the other.

There's a lingering residue of racism and phrenology to the hat-brim line too. The line was originally called the Hutkrempenregel, and it was first described during the later years of the European forensics boom in 1919. Still. When there's nothing but bone to examine, you have to consider everything, and there's a logic to the hat-brim rule that holds up. Recent reviews of the rule suggest that when a person falls from a standing position without any other blows interfering, the injuries will fall below that line.[1] Basically, the top of your head isn't at risk in most unprovoked falls.

The parietal area forming the dome and the load-bearing walls around the brain are an attractive target for a weapon, though. Even back when our most sophisticated weapon was a very pointy rock, it was easy enough to lift our arms and crack that bulbous egg.

Cranium 17 had two such cracks just above the hat-brim line. The fractured skull was recovered from the bottom of the Spanish archaeological site La Sima de los Huesos—literally, "the pit of bones." In 2015, the archaeological team at the Sima, led by Dr. Noemi Sala, used imaging, trajectory analysis, and the hat-brim-line rule (among other methods) in a *PLOS One* article to describe the holes in Cranium 17 as "the earliest evidence of lethal interpersonal violence in the hominin fossil record."[2]

In other words, a murder.

Because the owner of Cranium 17 died approximately 430,000 years ago, the facts of their life and death are frustratingly vague; we don't even know exactly what species this individual belonged to. But the circumstances of their death have a clarity that calls out to the violence of our world today. We see two fractures, identical in angle and shape, and know that the same branch didn't fall on their head twice. As the vibrant colors and stylish lines of nearby cave paintings bring a reminder of the indefatigability of human creativity, this battered head evokes the longevity of our need to kill each other.

Axes were not made for murder, but they were made in part for violence. Prehistoric hand axes were shaped to break down wood, chop up vegetables, and fatally pierce the flesh of animals. They were treasured for their utility in carving food and shelter from the unforgiving wilderness, but they were sharp in the same way a flame is hot. A fire needs to be hotter than we can stand so we can cook food and not die from the cold, and for that reason we tolerate the danger of fire. An axe is sharp because the danger is necessary; it is a flame made durable in stone, iron, steel.

Cranium 17's killer surely did not custom-knap a hand axe just to kill them. Perhaps they were already holding it, the fatal conflict rising while they crushed bones to extract marrow from animals or skinned them for pelts. Or they could have picked one up that was just lying around when murderous intent arose. Because a hand axe was always around, even in the chaos of a mass grave.

La Sima de los Huesos—the Pit of Bones—is a grave. Today it is also an excavation site, one of many in the Sierra de Atapuerca. The mountain range is just west of Burgos, where the researchers who work at the Atapuerca archaeological site stay near comparatively less ancient palaces and churches and enjoy blood sausages and other Castilian cuisine. The bones were first noticed when a mining company dug a rail trench in the foothills, abandoned in 1911. It would take another century for archaeologists to realize that the bones embedded in the limestone and speleothem and guano dated back a half million years.

There are a number of sites of early human activity throughout northern Spain, perhaps most famously the vivid bison drawn in ochre and charcoal thirty-five thousand years ago at Altamira, seventy miles north of Burgos. Atapuerca attracted and preserved an astounding concentration of evidence of prehistoric communities at many different points throughout the Pleistocene: the Sima del Elefante held a 1.3-million-year-old jawbone that proved to be evidence of the earliest of any species of human activity in Europe, and the Gran Dolina contained human bones that looked to have been cooked and gnawed upon around 780,000 years ago. The area was not continuously occupied—these were nomadic communities, moving in and making the mountains home for a few months or maybe years, and then moving on when the weather or the flora or the fauna changed. And as humans moved back and forth on an unfathomably long timescale, we changed not just our habits and tools but the shape of our heads and the length of our bones.

There's no consensus on to what species the bones in the Sima belong. Like a debate over what degree removed you are from a cousin you don't see often, the question seems to revolve around exactly how related they are to Neanderthals and modern humans. Cranium 17 and the rest of the occupants of the Pit of Bones are not quite Neanderthals because they are a little too small of brain. The people in the Sima had not much in the way of the occipital chignon—the bun at the back of your head that keeps the back brim of the hat in place. For a while—and in some corners still—the people of the Pit of Bones were called *Homo heidelbergensis*, but the very classification seems to just make it more complicated. Like, did *heidelbergensis* give rise to Neanderthals? Are they ancestors to that and other Homos, *floresiensis* and *erectus* and *sapiens*? Or were they just another branch on the family tree that ended?

There's not really a good answer. "It's kind of a mess taxonomically in the field right now," Dr. Rolf Quam, coauthor of the 2015 study that classified Cranium 17 as a potential murder victim, said to me in a 2020 interview. "[If you ask] three different anthropologists who work in the field what [*heidelbergensis*] is, and which specimens go in, you get three different answers."

Heidelbergensis is "sort of a wastebasket category."[3] Basically, any human remains older than 200,000 years and younger than 800,000 years that aren't clearly Neanderthal, aren't clearly Homo erectus, and aren't clearly modern humans might get called *Homo heidelbergensis*.

Despite their differences—brains a couple hundred cubic centimeters smaller, stronger fingers, enviably thick eyebrows—these were people, whatever name we use. "These are not half human, half chimp," said Dr. Quam. "If you saw one of these people walking around, and they were [in] regular clothing, you might not even notice them." They were as smart and adaptive and innovative as us, surviving with little more than their community, their wits, and a few basic tools.

The remains of as many as twenty-nine people reside in the Pit of Bones. Dr. Quam imagined that the community they came from was an extended clan of a couple hundred people, sharing the burial tradition as they did hand-axe technology. There are many places, of course, where a lot of bones are collected in one place, but they're usually inhabited sites, where people obviously gathered and worked and ate and slept—leaving behind hearths, or animal bones, or tools and the little pieces of flint carved away. Or just ash. There are other sites like that in the karst system—indications that they slept there, hung out there, probably venturing out to less-well-preserved open-air sites. They were "around and doing things," as Dr. Quam put it.

But not the Sima. It was inaccessible then and now, tunnels and caverns tucked deep beneath the rugged surface of the low-slung foothills. The Sima site is less than ten miles from Burgos, but over the brushy terrain it takes a good half hour to drive there, plus a half hour spent navigating the karst system, passing many other excavations of slightly later settlements. All-day excavations are impossible because of the lack of oxygen.

Once they're at the Sima, archaeologists are rewarded with the promise of the finds that might never come for other researchers. At many sites, bones get smashed "like a pancake," but in the Sima they are "undistorted." Researchers have even managed to get DNA from some of the remains. The Sima "is one of the richest hominen-bearing sites in the world. So, you know, every time you go down there, you're pretty much guaranteed you're gonna find human fossils," Quam said.[4]

And not much other than human fossils. Dr. Sala conclusively ruled out "geological causes, carnivore activity, and accidental falls" as a reason for the accumulation at the site. It's not the only cave in the system, but it's the only one with this many human bones and little else in one layer. The only way that many bodies could get down there is if they fell, or were thrown. There are too many bodies, and it's too remote, for it to just be a fall, or a random tragedy, or even ritual murder.

"The only access to the cavity is down the vertical shaft and there is no way out from the site once you are at the bottom of the shaft," said Dr. Sala. The Sima "is a place for the dead."

There is only one piece of evidence of a human settlement on the site: a hand axe called Excalibur. Found in 1998, the 500,000-year-old axe is made of ocher-veined pink quartzite. It is about six inches long, just under four inches wide, and more than two inches thick. It is classified as a biface because both sides were knapped (chipped away) to sharpen an edge. One side is flat, and the other side carries most of those two inches referred to above. Many similar axes abound at other sites in the karst system at Atapuerca, but those are clearly former butchering sites. There is no other "lithic waste"[5]—discarded stone tools—at the Sima. Because of its singularity, because it's the only nonsense of daily life that made it into the pit of bones, because it's so beautiful—the argument is that the axe is evidence of funerary practices. Just as you wouldn't bury your uncle with his beloved laptop, tools and other practical things are usually not interpreted as evidence of prehistoric grave decoration, the way beads might be. Laying nonperishable objects to rest with our loved ones is the same as any other decoration: it's about putting something pretty in there with them, to make it nice.

There are a lot of hand axes I find beautiful—there's something lacelike about the facets of each one, a textured delicacy to the chipped edges—and Excalibur is special. The saturated colors! The finely knapped lines! The intense ridges! It's very old and exquisitely honed, delicate and yet obviously made to endure.

I wouldn't call it "sexy," but some archaeologists would. In 1999, Marek Kohn and Steven Mithen posited that a great hand axe would have been an indicator for suitability as a mate, because making a hand axe is difficult and takes a lot of desirable qualities: Knowledge of local resources, like where to find the right rocks. The physical prowess required for the percussive technique of knapping (hitting or throwing rocks against other rocks to get them into the right shape). Ability to work with others, since hand axes may have been shaped from flats of stone prepared by others for "more standardized shape, thickness, and size" so they could be formed to better suit the intended tasks.[6] The intangible benefits of good character may have mattered even then—possession of an axe could serve as proof of the ability to plan and coordinate and possess the determination to actually finish the thing (I'm imagining a lot of half-made axes abandoned like novels). The symmetry and power of these tools act as a sort of plumage.

Hand axes like Excalibur were not axes in the sense that the term "axe murder" invokes. In violence or firewood production or the cooking process, a person's arms would have to do the work of the nonexistent handle in applying force and leverage behind the sharp edge (though some hand axes had wood or bone handles, wedged or attached with twine, as early as 500,000 BCE). Even without a handle, Excalibur is still dangerous, pointy and sharp, capable of stabbing a skull. Her edge comes to a single point, a little like a pickaxe or the sword she is named for. Really, Excalibur has as much in common with a noble sword as it does with an executioner's axe.

Dr. Sala mentioned that the weapon could be something else. "We know only that a blunt object was used," she told me in an email, mentioning a spear or wooden object, a "tool of standardized size and shape." Hand axes this early were sort of stem cells for future tools and weaponry. The stabbing tip, the whetted edge, the crushing heft: in this form they were ready to become many things. A broken axe might become the tip of a spear. The oldest axes were, in function, already knives, planers, swords.

Even in the Pleistocene, solid technology could be discarded in novelty; by the time of Cranium 17, the biface axe was already ancient. In some cultures it was obsolete. There's a marked lack of large axes in central Europe, replaced by lighter tools. So maybe Excalibur was just thrown away like a bulky old cell phone.

Perhaps it's sentimental of me, but I don't think that's why it was in the Sima. With its seductive symmetry, with that many uses, with that level of time investment, why would you throw it away? When objects are lovely and scarce and hard to hold on to, when objects of beauty are mostly fleeting and the province of nature, they become all the more precious. Giving Excalibur up would not be nothing. It's a loss to answer a loss.

Or perhaps to assuage guilt. Dr. Sala has found that many of the skulls in the Sima might have belonged to murder victims, another twist in a mystery that will only offer more mysteries. There's a funny paradox in how differently murder mysteries are handled when they're ancient as opposed to just old. When we look for answers about murders from a hundred years ago—*old* murders—we have no hope for answers truly beyond the bounds of reasonable doubt. Even in the true crimes that have been most thoroughly examined from every potential angle, like the case of Lizzie Borden, we're not working with blood samples or fingerprints. They're questions of storytelling and research and logic, but not of science.

But when we move the timeline so far back that it becomes almost impossible for us to comprehend, we have nothing but science. DNA cannot reveal a culprit by name but it can connect shards of bones. Measurements and advanced imaging and numbers that have been quadruple-checked—these are what we have in the absence of information about who these people were and why they might have killed each other.

The owner of Cranium 17 had many of the same flaws and strengths and challenges experienced by you and me. They were generous or forgetful or witty. They probably had projects they cared about and family they wanted to care for. They had conflicts, one of which probably turned deadly. But we don't have any of the specifics.

The narrative we have is in bone. There's a certainty to it that releases us from the painful catharsis of empathy we seek in other stories of murder. The act of burial, though, of being cast into the abyss—that is more discernably human. The grief of the not-quite-Neanderthal's surviving family brought them to the edge of a cavern to seek solace in putting their loved one to rest. They hoped for peace after a terrible end.

There are, arguably, good reasons to strike someone twice in the head with a sharp rock. Tyranny, slavery, abuse—sometimes a person must defend their life and freedom with arms. Perhaps Cranium 17 was guilty of one of those good reasons. Maybe they deserved it.

Violence is an expression of power, even if the stakes are incomprehensible to our eyes. But the two identical strikes to the head are not what is writ large across the parietal bones of human history. The preservation of the fractures in Cranium 17 and their neighbors was as intentional as their death, as careful as Excalibur's knapping. Care and attention that resonated through hundreds of thousands of years, discovered again and made novel by reverently rigorous inquiry. The skulls take the form they lost from the axe, from the burial; their reconstruction shapes new understanding for the depth of our human need to express violence and grief.

Type G-VII Egyptian Battleaxe

1550 BCE, Thebes, Egypt

One of the Egyptian hieroglyphs for the word "enemy" was a man with an axe in his head. By the time the pyramids took shape, there were hammers, saws, daggers—many other tools and weapons as common as the axe. But the axe was only gaining power.

When axes began to be consistently attached to handles between 12,000 and 7,000 BCE, they had not just more force but height—especially while held by a soldier on horseback. The sharp, dangerous part at the top was a fearsome thing for all to see. Shiny too, after metalwork began as early as

6,500 BCE. Many functional axes throughout the Egyptian period were still stone, especially since the axe was not high-status. Early Egyptian battleaxes were functionally the same as woodcutter's axes—a half moon perforated so that leather cords could be threaded through it and attached to lugs—only metal instead of stone. The malleability of copper and bronze alloys made shaping the objects much easier, refining their efficiency in the violence of the battlefield. And the rarity of the ore made the tools all the more valuable, precious. Worth showing off.

Technology often escalates conflict. We show off our new toys to make ourselves feel fancy and powerful, to impress the members of our own set, to intimidate our enemy. But the thing about boasting is that it also encourages imitation. An axe with a particularly dangerous point on the heel or toe of the cutting edge will be noticed and show up on the other side within just a few battles. Knockoffs often beat innovators at their own game.

Egyptians often took the weapons of their enemies for study. The Hyksos to the north and the Kerman culture in Nubia to the south already had technology that was downright transformative, like chariots. Even armor was a brand-new concept for Egyptians in the Middle Kingdom—as wild as it sounds, Egyptians basically fought naked, so an axe didn't need to be too sophisticated to find its way to harm. The Hyksos method of socketing their axes was new to Egypt—they fit the handle through a hole in the axehead, rather than fitting the butt of the blade through the wood or lashing it. Kerman warriors, long essential to the Theban military, forged axe blades that were longer, with a flared end.

In the Middle Kingdom, the epsilon blade—one long sharp and solid curve with three tangs fitting into slots in the handle—provided many opportunities to cut into enemy flesh. But as the Egyptians started to clothe themselves before they went to battle, they chose a different kind of axe, influenced by Kerman and Hyksos trade. These new axe heads were long, the cheek curving in and out like an hourglass, the blade turning back out into a flared shape. Hooks, almost, which struck through and tore apart armor and were particularly "conducive to severing heads."[1]

TWO

THE SMITED KING

Five hundred years before Seqenenre Tao died from battle-axe wounds, Thebes's ascent to capital of ancient Egypt was christened with a mass grave: sixty bodies, experienced soldiers of Lower Egypt, killed in a civil war by arrows and daggers and simple clubs. Their crushed skulls were left out to be eaten by vultures before being wrapped in linen with no other preservation. Effective propaganda for the brutal power of the new pharaoh Mentuhotep II, who had been just a regional king until he reunited the two lands of Egypt and began the eleventh dynasty in the mid-twentieth century BCE.

So the Middle Kingdom was brutally begun, one of the most glorious times in Ancient Egyptian history. Thebes was all of its greatest qualities wrapped into one city: the political, religious, cultural center of Egypt. Until a vizier named Amenemhat I took the top job from Mentuhotep IV, maybe in a coup, and moved the capital closer to Memphis.

In the eleventh dynasty, the new royal family doubled down on internal security and border control. But the tighter the hold a state has on its people, the greater the power vacuum it creates when it finally loses its grip.

Fifty pharaohs shuffled through power in the 150 years after the eleventh dynasty died with the woman king Sobekneferu in the early eighteenth century. These were elder statesmen who felt that they had waited their turn and now wanted their glory—all while making no more impact than your average William Henry Harrison. Viziers kept the country running, but Egyptians could feel the relief of lower expectations for kissing the king's ass. People stopped wedging mentions of the current pharaoh into their funerary monuments. Even the frequent new kings cared less about the majesty and dignity of the royal tradition; the pharaoh Sobekhotep III, circa 1680, bragged about his commoner parents, branding himself as a humble outsider, a breath of fresh air, a contrast to those fat cats down in Memphis. But he didn't last long either. Egypt's border security began to lose its grip as the royal court slid into a gerontocracy shuffle. Unpaid bills started to pile up, a betrayal of the Egyptian veneration of order. When the region of Avaris (then called Hutwaret) declared independence, the central government could barely muster a response. But the new state didn't last long either. Everything was a mess.

And then the Hyksos seized the moment.

It's hard to say much of anything definitive about who the Hyksos were. They weren't compulsive notetakers like the Egyptians, nor were they around long enough to accumulate a ton of monuments to themselves—they were active for about a hundred years, from the mid-seventeenth to the mid-sixteenth century. Hyksos literally means rulers of foreign lands; over a thousand years later, Egyptian historian Mantheo called them "shepherd kings" (shepherding was not a high-status profession in Egypt). They had roots in coastal Lebanon. To the xenophobic Egyptians, the "Semitic-speaking coastal elite" (as Toby Wilkinson described the Hyksos in his book *The Rise and Fall of Ancient Egypt*), were among the "miserable Asiatic."[1]

Egyptians revered their former glories; with a view like the pyramids, who wouldn't want to look back? But the Hyksos had no such fixations on the past. As the top-heavy Egyptian administration toppled, the Hyksos moved forward, focusing on practical issues, keeping the supply chain going. Lower Egypt was in chaos, and the Hyksos were the people trying to do something about it, trying to stop bad things from getting worse while the administrators went through a king a week. And for over a century, they ruled the Nile Delta and more, ruling Thebes for a short period.

The Hyksos were not tyrannical. There was some degree of cooperation between the Hyksos and the locals—Egyptians were able to own and profit from their lands for agriculture and animal husbandry, for one thing. When you weren't facing their newfangled chariots, the Hyksos weren't that bad to deal with. And the Hyksos weren't the only rivals for the red-and-white crowns of Egypt in the mid-fifteenth century; the Kush region of Nubia had a coalition of fierce warriors ready for the fray.

Some Thebans wanted to play it safe and simply adapt to the Hyksos presence in Egypt. But others in Thebes believed they were the true Egyptians, and that they should rightfully own the country. Avaris was becoming a thriving trade port with a distinctly Asiatic culture, a hub of fresh ideas and heterogenous energy. Egypt was in many ways a melting pot, and these Thebans didn't like that. They wanted a *real* Egyptian on the throne.

Seqenenre Tao came to the throne shining with the promise of a future that looked more like Thebans' glorious past. Tao didn't rely on simply riling up the base and going into a chaotic war—he made orderly choices in fortifying the area around Thebes. Nor was he a coward. When the time came for battle, he did not stay in his new strongholds but joined the front lines. He was easy to believe in, not just for his lineage or his bona fides in planning and executing war, but for the same reasons politicians succeed today. He had the right look—tall, handsome, great hair—and a formidable wife, Ahhotep. The reunified and triumphant Egypt that he promised would come to pass.

But he would not live to see it.

For many years, Seqenenre Tao's death was almost as much of a mystery as Cranium 17's. Because he was found with the likes of Ramses II in 1881, it took years before anyone figured out who he was. Everyone was so busy with the other mummies in his cache—like Tao's son Ahmose and the Thutmoses I, II, and III—that it was five years before anyone actually took him out of his coffin and took a good look at him. Or, well, a good sniff of him. Most mummies smell pretty good for being dead that long—not great, but better than decomposition. Definitely musty, but stuffed with oils and perfume, so it's more like "stale beef jerky" than anything truly stomach-turning.[2]

Seqenenre Tao *stank*. His "spicy"[3] sort of foul smell was the first clue that something was different about him when he was finally examined by Gaston Maspero in 1886. And the body was in no better shape than his shroud: the head and bones were "loose and misplaced."[4] As Maspero unrolled the greasy, worm-infested linen ribbons, Seqenenre brought him layer after layer of horror unbefitting the careful standards of royal Egyptian embalming: mutilated head caked with brain matter, beetle larvae in the once-glossy hair, a head completely separated from his body.

Maspero imagined an axe from the very beginning. But for more than a century, researchers poked at Tao's royal anomalies, suggesting different theories of his death. Since the wounds came from above and none were defensive, he could have been assassinated while asleep. Maybe with a dagger! For others, the position and multiplicity of the wounds suggested a battlefield melee: chariot-borne, axe-wielding Hyksos soldiers descending upon Seqenenre Tao like a sudden plague.

It wasn't until Garry Shaw began looking at the case that the story came together, the narrative well before the science: "I'm not a technology specialist. I'm not a mummification specialist," he told me via Zoom. "I class myself as more of a historian of ancient Egypt. And so I was just synthesizing all the information I could possibly find and pulling it together."

By understanding the political context of the king, he was able to envision his death in a way later verified by physical evidence like imaging. Not every scholar agrees with his interpretation, but a consensus of sorts settled around Shaw's version of events.

The exact cause of the fatal conflict and many other precise details like its scale and length are gone. What we do know is that the "prince of the southern realm"[5] braved the battlefield himself in 1533 BCE. Probably not on the front lines, but he was with his men in Hyksos territory when he was captured. His army was not around him when he met his end. When he was executed, Seqenenre Tao was a prisoner in the realm of his enemies, the rulers of foreign lands that he thought were still his. His hands were tied behind him, and he was forced to face his final defeat on his knees before a power he could not displace.

Likely first was a wound just above the hat-brim line, high on his forehead. The axe was the newfangled kind with the splayed blade, wielded by a high-ranking member of the Hyksos army—perhaps Apepi, who appears with Tao in an incomplete piece of Egyptian literature in which the Hyksos nomarch complains about hippos (often a slur of sorts in Egypt, though in which direction I'm unclear) to the Thebes ruler.

Seqenenre Tao fell to the ground as his Hyksos assailant continued their blows, cutting his right eyebrow and left cheek. They used the butt of the axe or perhaps a mace to smash the king's nose. Someone else jumped in then, spearing his ear so deeply that the sharp thin tip of the weapon reached Tao's foramen magnum—a small opening in the occipital sphere where the spinal cord lives.

The pharaoh was definitely dead after that. Someone kicked him to his side, then stabbed the right side of his skull several times. Then they left him there to decompose. No linen or anything, less care and reverence than even the sixty soldiers killed by Mentuhotep five hundred years before.

The choice of the battleaxe was probably not an active consideration but an instinctive practicality. The axe was a bit lower status, a bit more common than a fine dagger or knife or sword, but it was the instrument of that war. If they'd returned with a living captive king, they might have used something other than an axe to add to the ritual, but the Hyksos won northern Egypt when they chose to address problems urgently instead of discussing them with viziers until they became catastrophes. They chose to take their moment when it came instead of taking their time.

Execution and wartime killings can be abusive, of course. But the Theban king's death seems to be on the level. His death was violent, terrible, and they did literally leave him there to degrade. But there's no evidence of prolonged torture or bodily desecration. His killers were not his hosts or his guests, violating important laws of hospitality to gain an advantage. It was an act of war, not a secretive murder or a ritual show of power. They were enemies, doing what enemies do.

Though Seqenenre Tao's story ends there, the story of the Egyptian battleaxe in this period of transition does not. The mummies of Tao's wife and sons did not bear the evidence of the axe seen in his destroyed head. But in their own reigns, in their less calamitous deaths, they reached for the axe to demonstrate that they held the power to smite slaves, citizens, and kings.

The nature of Seqenenre Tao's relationship to the next pharaoh, Kamose, is unclear, because Egyptian royal family trees are extremely difficult to figure out in the best of circumstances. Ahhotep was an ideal queen because she was Tao's sister, after all. Son, brother, whatever, Kamose certainly had the kind of chip on his shoulder you'd expect from an underdog substitute king. He liked to see himself as someone who went against the entrenched wisdom of his snooty advisors, who wanted to stick with the three-state status quo. Kamose's kingly ego wasn't satisfied with quietly thriving. His wish was to revive Egypt.

After blasting Hyksos-supporting towns, he intercepted a letter from the Hyksos pharaoh Apepi to the ruler of the Kush offering to split Egypt between them. He sent the letter back with a message of his own: "I will not leave you alone, I will not let you walk the earth without my bearing down upon you."[6]

This was no empty threat. Soon he was in Avaris, within sight of the royal compound, their women peeping out "like baby mice inside their holes." He raped them, and cut down Apepi's trees, and took his precious goods: "gold, lapis lazuli, silver, turquoise, bronze axes without number."[7]

But when Kamose died a couple of years later, he wasn't buried like a conquering king. There was so little drama around his death that it feels like a rebuke; Kamose was stuffed "in a grave suitable for a private person rather than one made with a king in mind."[8]

As soon as Ahmose reached adulthood, he hit Lower Egypt with full strength, taking back first former capital Memphis, then Ra's cult center Iunu, and finally, after two extended aquatic assaults, the Hyksos stronghold Avaris. The army of central Egypt plundered the city, killed the Hyksos, and then hunted down and killed anyone they could find who fled. Then he took over a little of Palestine too, just for good measure. Aphosis was defeated, and the Hyksos rule of Egypt was over.

Holding the center in the midst of all this Kamose and Ahmose chaos were the queens. In the years between Kamose's death and Ahmose's assumption of the throne, Seqenenre's widow Ahhotep led the country and kept it stable, maintaining trade and military ties and keeping out the Hyksos in Lower Egypt and the Kerman in Upper Egypt. Seqenenre's mother Tetisheri, Ahhotep, and Ahmose's queen Ahmose-Nefertari were as powerful as pharaohs, their long lives accumulating public glories. Ahmose's esteem for his mother Ahhotep was recorded in one stone monument, especially her accomplishments in having "pacified Upper Egypt and subdued its rebels."[9]

Between this and the many axes in her makeshift tomb, you can see why early Egyptologists thought of Ahhotep as a warrior queen. Her mummified remains were discovered in 1859, but her sarcophagus and earthly remains were confusing because she was moved thousands of years ago, from her original resting place, to avoid grave robbery. Her necklace of golden flies was instantly iconic (it was reproduced in Elizabeth Taylor's *Cleopatra*) and was the basis for the idea that the golden flies represented tokens of military accomplishment. More recent research suggests that these flies were not badges of military honor or persistence but personal totems with a variety of meanings—beauty, health, nature, life, death. Flies became not a persistent annoyance but the manifestation of the things we value most, made into wood, bone, metal ornaments to become an element in one's personal style.

Ahhotep had a total of twelve axes buried with her. Eight were little things, three and a half inches or so, made of gold and silver. They are plain except for a little embossing in imitation of cords lashing the head to the handle. Possibly they were game counters, like the thimble in Monopoly.

The big axes seem to be from the reigns of Kamose and Ahmose, because keeping old stuff at your mom's house is an ancient practice. They are all of the type that killed Seqenenre, hourglass-shaped with a broad cutting edge. Two praise the name of Kamose, son of Ra, in hieroglyphs—identical blades but only one with a handle, made of horn. Both are bronze, slightly asymmetrical, with some pitting at the lugs and frayed at the cutting edge—signs of use. These were functional objects. Whether they were used to make ceremonial furniture or cut ceremonial wood or cut off heads ceremoniously is unknown, but they were tools in use and not just something to behold.

When the state was weak in Egypt, functional axes and daggers were much more common in mortuary contexts. When Egypt was strong, the axes in graves were much weaker: they were abstractions, symbols of power. Tokens. Ahhotep was the power behind a weak and a strong state in her long life. She got the country through the failures of Seqenenre and Kamose into the victories of Ahmose, and so the life-size axes of Ahmose in the tomb of Ahhotep are not fit for battle: their waists Barbie-narrow, their gold and silver too soft to sever. They are just for show.

But what a show! There are two of these beautiful axes, but we only need to talk about the fancier one. Shining gold offsets images inlaid in lapis lazuli, turquoise, carnelian. There is a man kneeling with palm branches to represent eternity, a vulture and a cobra both wearing the two crowns of a united Egypt, a crouching sphinx. On the other side is Ahmose's names, and below them a scene of royal violence.

Ahmose stands tall, holding the frizzy hair of an enemy. Hair shape was often a way to mark the other in ancient Egypt; they were all about style, but what culture isn't? This doomed person was probably not a Hyksos warrior, who were signified by a distinctive mushroomy cut, but a Nubian or perhaps a traitor Egyptian. Ahmose wears his royal kilt and holds a khopesh. A khopesh is the most uniquely Egyptian weapon, somewhere between a battleaxe and a sword. It is one long piece of sharp metal, curved into a sort of C shape, which is most often compared to a sickle but reminds me more of an epsilon-shaped axe.

The pharaoh is seen in the act of smiting. As he sinks the khopesh into the frizzy head, he holds his victim's elbow to stop them from fleeing. They are crouching, or kneeling perhaps. It seems a scene of war and not ceremony, like Seqenenre Tao's end.

Below the smiting is a griffin. A griffin is a Greek symbol, dating back a few hundred years before the axe was cast. The Hyksos occupation had the side effect of opening up the Levant to Egypt, and Ahhotep's stela (commemorative stone slab) records among her other accomplishments that she was the "mistress of the shores of everything-around-the-islands."[10] Islands where the griffin was an emblem of women holding power.

The queen's finest axe outshines all her jewelry, the many intricately beaded and etched collars, bracelets, earrings. As Egypt became more stable, there was more time for ornamentation and feminine touches; green ribbon and golden fish were some of the rewards men vied for in the early New Kingdom. Ahhotep's steady leadership in the years following Kamose's death could certainly give greater credibility to the women in the royal family seeking their own power. Within a century, Hatshepsut would follow her example.

Ahhotep may not have been a warrior queen; there is no solid evidence that she frayed that bronze axe with the horn handle. But it says it right on the stela: she ruled Egypt. And wartime or peacetime, king or queen, one land or two or many more: exercising power is no benign affair with soft rounded corners and kind words. It is a thing of secrets and recriminations and cutting edges.

Even if Ahhotep didn't hold the axe herself, she held its power, kept it safe and sharp for her sons. Her totems were individually harmless, even charming, but they were representative of a real and terrible power, a sharp and heavy tool ready to chop down anyone in the name of the power of the two crowns.

Yue

1200 BCE, Yinxu, China

The earliest Chinese word for king, "wang," may be translated as "the big man." But there's another translation that made me do an excited little clap when I found it: "the man with the axe"—which, as my source helpfully clarifies, is "used for chopping off heads."[1]

The primary axe of ancient China, called the yue, is a lot different than the axes in Ahhotep's tomb. Some yues are actively menacing, with eyes and a horrible rictus grin cut out of the bronze, square-toothed jack-o'-lantern style. The yue is thin and heavy, big enough and recognizable enough from a distance for it to be tied to the top of a tall pole and used as a standard for the military to follow. These were bespoke weapons, each tailored to the particular danger of the carrier. The process of design and casting would take a lot of time and resources, creating the kind of object meant only for the highest of high, kings and generals. And maybe a queen.

For a time after we began to understand metal, the axe remained the symbol of power and military prowess. As we started to record our own existence on papyrus and clay tablets and bits of bone, we used axes to communicate something more primal. Far from a tool taken for granted, an axe raised high on a pole was a provocation, an expression of might. China was one of the greatest producers of bronze in the ancient world, and in the Shang dynasty the axe was its totem: weapon of the executioner and the battlefield, a canvas on which to show off artistic and technical excellence, and a flag for warriors to follow into combat. Shang dynasty warriors didn't use battleaxes in their hand-to-hand combat. The yue was more authoritative than that. It was only "bestowed upon generals who had the right and power to levy war."[2]

Kings displayed the axe to remind their subjects of the consequences of disobedience. Executions were not reserved for lawbreakers. Enslaved people were frequently executed as a part of ritualistic violence. State violence performed with an axe is not necessarily the same as axe murder as an act of interpersonal violence. Which is not to say that they're any better; few righteous processes end with an axe in the head.

But in many cases they are decisions made not by one person but by the community. War and execution and even human sacrifice have to consider customs, ritual, circumstance.

The Shang dynasty king Wu Ding hung an enormous axe with the face of an eager, terrifying beast behind him on the wall when he met with his subjects. This wasn't a functional thing, this yue. It was certainly sharp enough to chop off a head, but it was massive. The proportions are like those of a fantasy axe in a video game, technically capable but simply too huge to pose a practical threat. Yet it was certainly a threat. The axe was power itself.

The Shang axe embodied the fear and might that underpinned Wu Ding's relationship to his subjects. The axe and the menace of execution, the implicit force that made the king's will be done, were so internalized that he didn't need to actually wield it: "The axe went straight from the wall to their hearts."[3]

THREE

CASCADE OF BLOOD

The queen Fu Hao was buried with four axes, but she began her life as a horse girl. She was the beloved scion of the Hongshan culture of horse breeders, an ideal princess bride for Wu Ding. He had a habit of marrying women from tribes and communities at the edge of his kingdom, presaging the harem of the rear palace system with his "living symbols of Shang diplomacy."[4] Like his other wives, she was not just an ornament from his travels but a formidable ally who spent her childhood in military training to prepare for a life of power.

The Zi dynasty must have seen her marriage to Wu Ding, then a prince, with great pride and love leavened by the sorrow of losing her. Wine vessels from her maternal line that may date back to before her wedding address her as "the smart and lovely girl."[5]

Not long after her wedding to Wu Ding, her father-in-law the king died, and the family was obligated to formally mourn and abstain from courtly life for three years. The young couple used the time to get to know their country. Encountering other communities on a nonwarfare basis could have been a rare opportunity even for these most privileged leaders, and the couple took the time to learn about the local landscape: the crops, the irrigation methods, the terrain. It was a time of agricultural scarcity, and Fu Hao and her husband tried to mitigate the pressures of hungry times by holding meetings with local officials and the public.

Once the mourning period was over and the royal couple had returned to the capital city Yinxu, they encountered immediate military trouble in need of a creative solution. There was a boundary dispute at the northern borders, but all their best men were already deployed in the south to address a separate conflict. In this crisis, the new queen volunteered for service.

Wu Ding was conflicted. The Shang dynasty was certainly not a matrilineal society, like the Yangshao culture of the Neolithic era. But it wasn't as patriarchy fixated as the next society; only in the Zhou period did the axe become "clearly a symbol of male authority."[6] Fu Hao was raised with the entitlement of royalty and the assurance of military training, but her rise in the fighting ranks was far from predetermined. She held a unique position distinct from the courtly responsibilities of other wives, garnering the kind of respect enjoyed by few others in her milieu—women or men. Thanks to her high-status family of origin and the diplomatic experience of her royal honeymoon, she was becoming a political force of her own, and she was ready to turn soft power into sharp bronze.

When Fu Hao got her axe, she raised it high so her soldiers could follow behind her, and brought it down against threats to her empire. Sometimes she acted as the advance team for Wu Ding—finding lodgings, spying, tending to wounded soldiers, scouting the battlefield. Other times she was the boss. She led thirteen thousand soldiers in battle against the southeastern state of Jiang. Hundreds of miles to the north of Yinxu in modern-day Mongolia, Fu Hao was a key part of Wu Ding's campaigns. Perhaps her greatest triumph was in the southwestern state of Bafang: she snuck into Bafang to occupy the enemy's path of retreat; when Wu Ding attacked them, she confronted them as they escaped. The royal couple destroyed their enemy.

Fu Hao's life is attested to in oracle bones—shards of ox bone and turtle shell that were burned until they cracked and then interpreted for news of the future. The predictions—question, divined answer, and outcome—were written upon the bones, leaving some of the earliest writing in Chinese history. Fu Hao appears frequently in the bones, to mixed results. One of Fu Hao's pregnancies was incorrectly predicted to be a boy, which would have been a disappointment. But she was successful as ever at court, taking on senior roles in official duties and ruling over disputes among Wu Ding's other consorts. Her son the heir honored her well, giving gifts (but only after he checked with the oracle bones).

But soon after she returned to Yinxu, she got sick, and then her son did too. The prince died first, followed by Fu Hao, followed by as many as sixteen other enslaved people sacrificed to her afterlife.

When Fu Hao's tomb was discovered in 1976, it was a landmark in ancient Chinese archaeology. But what made her grave so significant wasn't really the axes. Most of the things in her tomb weren't unique, not the jade parrots or the beauty supplies or even the people buried alive with her. Her military service is deeply fascinating, but we'd have no idea about it if it weren't for the thing that truly sets her story apart from her royal peers.

Fu Hao's tomb is different because it was never robbed.

Fu Hao's unmolested crypt may be directly related to her secondary status. Wu Ding and his top wife Fu Jing were in the main tomb complex. Fu Hao's smaller tomb was apparently harder to find. Since her son predeceased her, she wasn't the heir's mother at the time she died. She's clearly an important figure, but the world of the Shang dynasty hereafter is deeply competitive. As powerful as she was on the battlefield, she was a woman. And even though she was the first wife chronologically, she did not die as the top woman—perhaps because of her early death. Still she was prepared to live the afterlife lavishly. There were two hundred bronze vessels for food and wine offerings, some from her life on earth and some from her mourners, many almost two feet tall, depicting unique and fanciful birds with dragons coming out of the back of their heads.

The tomb also held twenty-seven knives and over a

hundred other weapons. One jade dagger-axe had an inscription indicating that it was one of five daggers given as tribute by an outlying tribe. This isn't the sort of thing that showed up in the tombs of other consorts—this was Fu Hao flexing her martial power, not her marital power.

Fu Hao had four axes in her grave, two of them weighing over seventeen pounds. The most famous of the four features an expressionless bald face projecting from the center of the yue, menaced from both sides by stylized tigers. The others feature dragons and owls; all bear Fu Hao's name. The axes set her apart from other royal women, suggesting a power withheld from other consorts.

"Because the axe is present in her grave, that means that she wields the power over life and death," Dr. Keren Wang, a scholar of Chinese legal and intellectual history at Emory University, told me. "That means she can put people under trial and declare them guilty and execute them without approval of the king. And that's considered extremely significant power."

It's exhilarating to find queens of the past like Ahhotep and Fu Hao, to see women in power during eras in which women's contributions were usually erased. But the war-crime-level excess of Fu Hao's life and death were not a moral triumph. She lived her elite lifestyle on a wave of violence and enslavement embodied by rituals of human sacrifice. As many as sixteen people (and six dogs) around (and under) Fu Hao's tomb were a pretty modest offering by royal standards; the scale goes way up as you move up the royal order. Nine thousand people were killed in just one of Wu

Ding's ceremonies. In the case of Fu Hao's tomb, the form of sacrifice was called Renxun. Though these were mostly ancestral rites, they also served to "instill fear into the spectators, which ultimately helped to dissuade both internal dissent and attacks from outsiders."

Dr. Wang is a scholar of human sacrifice, and in our conversation he explained how modern sacrifices that have become mundane parts of daily life are connected to the bloody spectacles of centuries ago. The framework of a ritual conditions us to accept the unacceptable—even when the trappings of the ritual are no longer oracle bones and vestal virgins.

Triage is the modern human-sacrifice ritual I've thought of the most since. The process of quickly sorting out which patients need the most urgent care in hospitals means that some of the patients most in need of care will die so that time and energy will be spent instead on cases who stand a better chance of survival. As a ritual, the process imposes order on an inevitably tragic decision so that the way can be cleared to prevent others from dying unnecessarily. Another, less sympathetic human-sacrifice ritual practiced today is the commonplace decision to axe thousands of jobs in order to guarantee stockholder happiness, the kind of bloodless brutality that we cannot accept and cannot resist.

The collective stakes of human sacrifice were crucial in strengthening the royal court in the Shang dynasty, an early example of centralized government and social stratification: the rituals reflected and reinforced the norms of the day. They were as much political theater as they were religious ceremony.

One of Wu Ding's rituals, called forth to end a famine, struck me as Kubrickian in its cinematic use of language: "a cascade of blood," as Dr. Wang explained it. The ritual was structured on a step pyramid, with sacrificial victims numbering in the hundreds or thousands stationed atop it. These men were decapitated, but the axe didn't stop there. Their torsos were chopped in half in order to generate the amount of blood needed for a waterfall effect. Some of the blood was also used for wine, but the main purpose of this flood of blood was to invoke rain, the axe recalling the thunder.

The people in the cascade may have known that they were headed for sacrifice, unlike the victims in Fu Hao's tomb. Her human sacrifices were caught unaware. "They thought they were coming back," said Dr. Wang. "They were just getting ready for the sarcophagus and getting all of the ritual objects set up. And then they turn around, and people are closing the lid."

Suffocation killed them, not the axe. But it was the threat of the axe hanging on the wall that empowered these rituals of sacrifice and slavery.

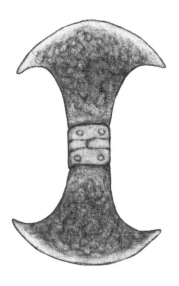

Pélekys (πέλεκυς)

520 BCE, Greece

In Herodotus, there's a very stirring speech from a Spartan soldier in the midst of the Persian war:

> *You know well how to be a slave but have not yet experienced freedom, nor have you felt whether it is sweet or not. But if you could try freedom, you would advise us to fight for it, and not only with spears, but with axes!*[1]

While a spear may be basic, it is, unlike an axe, definitively a weapon. In going beyond the spear to fight for their freedom, these soldiers are reaching past the obvious weapon to something more primitive—and perhaps more easily available not just to trained fighters but to other laborers. They are fighting for their freedom with every possible resource. And the axe they use to fight for their freedom is described as pélekys.

Later on in classical antiquity, the axe was a much more official emblem of might. Fasces were bound bundles of rods that sometimes had an axe protruding. The axe was often a double axe, also called a labrys or bipennis. Of course, actual fasces are no longer in use, but they are a lasting symbol of the brutality of state power. They are still in constant decorative use by governments obsessed with military might, whether or not they might be classified as fascist. You can find fasces in the ornate details of American life: the dime, the House chamber in the Capitol Building, bridges, flagpoles, gates everywhere.

Fasces were first hoisted in ancient Rome. Lictors were, like axes themselves, so omnipresent as to be below notice, a practical and symbolic extension of the state who accompanied magistrates everywhere. Lictors carried fasces; the axe was removed from the accompanying rods in situations where the magistrates did not have the power of execution. The rods and the axe were usually symbolic, but there were many situations where they did actually use the rods and axe to beat and execute condemned men—the kind of violence that inspires resistance.

FOUR

IN TRUTH, AN ENEMY AND A MAN OF VIOLENCE

To the richest man in the world, we are all potential fodder for human-sacrifice rituals. In 550 BCE, Croesus threatened genocide as a professional courtesy. The king of Lydia had just heard a prediction from the oracle of Delphi that left him feeling decidedly overconfident. When his ally Miltiades had a conflict with the people of the city of Lampsacus, Croesus was quick to threaten them with annihilation. Miltiades's power was secured. But soon after, Croesus's unimaginably wealthy empire fell to Persia. And a generation later, inhabitants of Lampsacus returned his favor with an axe in the tyrant Stesagoras's head.

Miltiades the elder and Stesagoras were both members of the Philaids, a famous Athenian family who relied on the largesse of Persian kings to maintain their status. His half brother Cimon was the winner of Olympic horse events, but was also known as Coalemos—"the nitwit" or "the simpleton." This probably wasn't so much because he was actually stupid as it was his exaggerated folksiness. He was a sophisticated man who projected an aw-shucks kind of air to offset the thick stench of aristocracy in his life, as David Stuttard explains in his book on the family, *Phoenix: A Father, a Son, and the Rise of Athens*.

One day during Croesus's reign, Miltiades was relaxing on his front porch in Athens when the Cheronese tribe walked by on their way to the Gallipoli isthmus. He could tell they were out-of-towners from their clothes and their spears, so he invited them in for food and entertainment. As it happened, according to Herodotus's telling of the story, this act of hospitality had been foretold.

The wandering tribe—part of the larger Thracian culture of Persia—had recently been informed by the oracle at Delphi that the next person who gave them hospitality should be their ruler. Not their king, or oligarch, or archon, as they were calling it in Athens. But their tyrant.

Tyrants could be almost beloved in the sixth century BCE. The word once meant a ruler who isn't exactly a king or an archon (chief magistrate) yet is very definitely in charge. Force was usually a part of it, but persuasion too. Miltiades was currently under the supervision of the tyrant Pisistratus. Pisistratus is remembered well, a ruthless but also effective leader who managed to maintain peace while guiding

Athens into maturity as a financial, political, and cultural center. Miltiades felt very much in Pisistratus's shadow, so Pisistratus encouraged him to form a Thracian outpost.

The whole story about Miltiades being invited to take over as king because of his generosity seems like a euphemism for a semi-forceful takeover, possibly backed by Croesus. The Athenians were mighty, but they prided themselves on fairness, hospitality, and of course democracy. Like colonialists always do, they thought of themselves as gifting their culture and language and customs to these rural, uncultured, sensual barbarians, the Cheronese. And if they had to shove another Thracian culture—the Lampsacus—out of the way, that was fine with them.

But the Athenians told the history, and so the account that survives says that the Cheronese welcomed the wealthy and generous tyrant. He quickly centralized and updated the pastoral, tribal infrastructure to resemble the bustling Athenian city-state. But there was resistance from outlying tribes. Miltiades soon built a wall (classic tyrant behavior) to defend the Gallipoli peninsula from outside forces and further Hellenize the culture of the land. This affected the important waterway, the Hellespont, which, then as now, was a major supplier of wheat to much of Greece and Persia.

It was a good place for an ambitious man like Miltiades to make some noise on behalf of Greece. Miltiades specifically targeted the Lampsacenes, the people of the thriving port city of Lampsacus just south of the Hellespont; defeating them would have given Athens control of much of the rest of the peninsula. But Lampsacus knew he was coming. Miltiades was captured upon arrival, with little recourse. So he had to call his rich buddy Croesus to bail him out.

Croesus sent the Lampsacenes a threat: if they did not release Miltiades, he would "wipe them out as if they were a pine tree."[2]

At first, the Lampsacenes didn't understand the threat. What leverage is a pine tree against a king? But then an elder understood: "The pine alone of all trees does not produce any new shoots once it has been chopped down, but it is utterly destroyed and gone forever."[3]

Croesus was telling Lampsacus that it was well within his power to totally flatten them. If he took an axe to them, they wouldn't survive.

The threat worked well for a generation. Even after Croesus died, Miltiades carried on leading the Cheronese, keeping peace during Persia's period of discord and building a good reputation among his people. But the Lampsacenian pine trees kept dropping cones. They grew into little trees, with institutional memory. Little trees growing under the shade of Attican pressure, under the threats and tensions of war. Little trees who grew ring by ring, remembering their embarrassment and seeking revenge. Little trees who had their own axes.

Over his tenure, Miltiades had become a popular figure to the Cheronese. But back in Athens, his brother Cimon the nitwit was pissing off his own tyrant. Pisistratus was constantly on the lookout for anyone who might eventually rival him, so he exiled Cimon and his two sons, Stesagoras and Miltiades the Younger. Cimon probably did something to get himself in trouble—one theory says that he was exiled for "ostentatious use of horses."[4]

Cimon bounced around for a while, Stesagoras and Miltiades the Younger in tow. David Stuttard suggests that Miltiades the elder did not host him in his Cheronese home because Miltiades didn't want to piss off Pisistratus, but it's easy to imagine Cimon, the lovable simpleton, charming his brother into housing him for long enough for young Stesagoras to get the lay of the land. Wherever they ended up, it was certainly in the luxury necessary to keep several horses groomed and to maintain dwellings decorated with fabulous art like a vase finely wrought in black figures against the red clay that depicts a teenage Stesagoras "draped in himation" (a special kind of wrapped cloak) and attending to one of those all-important horses.[5]

Ostentatious or no, Cimon was pretty good at using those horses. He dominated the 536 Olympics. He used the occasion of his second win at the 532 games to ease himself back into Pisistratus's good graces. Not only was the victory dedicated to the good tyrant, Cimon declared Pisistratus to be the *real* winner of the race! And so Cimon was allowed back into Athens, because sucking up works. But sucking up to a septuagenarian can be a risky prospect. The elderly Pisistratus died in 527, just a year after Cimon's final Olympic victory. The very fact that Cimon had recently made good with Pisistratus made Cimon a threat to the family once more—especially considering that the olive crown he won in the Olympics was still pretty fresh.

Pisistratus's son Hippias was the bad kind of tyrant. He murdered Cimon Coalemos the nitwit in the town hall, right in front of everyone. The circumstances of Cimon's death—ambushed in a government setting—neatly foreshadow his son's entrapment and murder in a government setting a decade later.

Stesagoras was gone by the time of his father's death, already in the Cheronese working with his uncle Miltiades, who would have been in his sixties by then. He had to do the advance work, because they no longer had Croesus to fall back on. There was ample precedent for a nephew inheriting the throne from his uncle, but the childless Miltiades proved to be a tough act to follow. Tyrants in Herodotus's accounts tended to die without issue, and apparently it "was a well-accepted truth in Classical Greece that tyrants had disordered sex-lives."[6] I hardly know enough about Stesagoras to comment on the order in his sex life or lack thereof, but he too would die childless.

When Miltiades died, Stesagoras kept the focus on horses by observing his death "in the customary manner, instituting a contest of horse races and gymnastics."[7] The Philiads cannot be faulted for their branding. And luckily for him, Thracians loved horses too. I'm sure the Lampsacenes probably also loved horses, because horse races were basically the soccer of the ancient world—everybody loved them.

Except the Lampsacenes weren't allowed to participate.

The tensions with the Lampsacenes over the Hellespont had not healed but festered. About a decade into Stesagoras's tyranny, the fight over the Hellespont continued to dominate. The tyrant convened a meeting of the war council. In attendance was a nameless member of the Lampsacenes. Stesagoras was assured that the man was a valuable deserter, but in truth he "was an enemy and a man of violence."[8] (Pausing among these scarce facts for a moment here to wonder: Who was tricking Stesagoras, and who was being tricked? Was the Lampsacenian man concealing his identity from everyone, or was there someone on the war council who knew what was going to happen? Herodotus doesn't say, so I can't either.)

In any case, the "hot-tempered foe" did not approach wildly.[9] He did not barge into the city hall like an axe murderer from a child's imagination, with a maul raised high like a lumberjack splitting a great pine. His true intentions were as well hidden as the weapon beneath his robes: an axe (in the original Greek, πέλεκυς, or "pélekys").

A hatchet is an axe, a maul is an axe, a pélekys is an axe. The exact kind of axe was not specified by Herodotus. Stuttard imagined a hand axe in his book, but the word πέλεκυς isn't specific. "Pélekys" means just an axe.

Writing in 2005, the scholar Lionel Scott interpreted the pélekys as specifically not a battle weapon, noting that the axe didn't have the implication of beheading that it would later on in the classical era. The weapon used to kill Stesagoras more strongly recalls the ritual sacrifice of animals, not man. And maybe "not man" is what the pélekys is about. "The implication is that the assassin was no soldier and killed Stesagoras with a woman's weapon," writes Scott.[10]

"Herodotus exaggerates most things," David Stuttard

told me in our conversation. Herodotus may have simply recorded the kind of weapon used. Or he may have used poetic license. Herodotus is considered our earliest historian, but he was no neutral observer. He wanted to put his stamp on events, make them a little grabbier. Croesus threatened to wipe out the Lampsacenes as if they were trees, and so when the Lampsacenes had the chance to get revenge, they used the tree's natural enemy. I think the axe was highlighted for literary reasons.

In many ways, Herodotus initiated the tradition of writing true stories down for both the sake of posterity and for the entertainment of a popular audience. Despite the former and because of the latter, he was also a bit of a bullshit artist, "history's sketchy uncle," as Evan Allgood described him in a *New Yorker* humor piece.[11] He was our earliest historian, this writer who invented many of the customs of the genre whose survival you, personally, contribute toward by reading this book. You're only reading this because you hope it's a good story. And as a sophisticated reader, you know that any good story is bound to take a couple of liberties. Not intentionally, of course. Or mostly not intentionally. Any time a story gets retold—in oral histories, by a scribe in training, during a scholarly translation, over a podcast, or dished in a gossipy post online—something's going to get just a little bit distorted. It's a part of making the story worth remembering.

When Herodotus tells you about a Persian king's death from a random leg wound, he's going to tell you that actually it's divine vengeance for the time that king killed an Egyptian bull god by stabbing him in the same place. When he wants you to remember a superb Corinthian musician named Arion, he'll describe how a dolphin saved Arion from pirates and carried him and his lyre to shore. He "strived for accuracy,"[12] but he was more concerned with understanding the world than getting every detail right.

The story of Stesagoras's death isn't reported by other sources, so we can't tell much about how faithfully Herodotus told this story. In some cases, he would discuss the stories he heard as they were told to him—for example, the story of Arion the dolphin rider came straight from the Corinthians, backed up by the people of Lesbos. But he doesn't give us much context for how he learned about this little incident. Stesagoras was a member of a famous family, who presided over a key area of Greek trade, but little was transmitted except for the way he was murdered.

But there's a lot about this era of Greek and especially Lydian history that is vague. Much of what we know about Lydian history that doesn't come from Herodotus or archaeology arrives via its intermixture with Greek language, culture, and mythology. Sometimes the disparate threads get a little kinky.

Croesus traced his ancestry back to Omphale, a Lydian queen, the daughter of a river god and the widow of a mountain god. Herakles—aka Hercules—was sold into bondage to her by Hermes. Perhaps they were married; the stories vary, but in their dynamic she was clearly on top. Herakles was compelled to do women's work like spinning and sometimes to wear Omphale's clothes. Which he seems to have been into. In their marriage, Herakles gave Omphale a powerful gift: a double axe, taken from the Amazonian queen Hippolyte.

The name for a pélekys with two cutting edges and no butt is labrys, a Lydian word more associated with Greece and the mythic Amazon warriors. The double-sided axe was a major object of worship in Crete, stamped all over the island on shrines, pillars, doorframes. The word "labryinth" is itself a reference to Cretans' "axe-fetish,"[13] as the labrys was associated with a deity that was not a father-god but a mother-god.

The symbolism of this particular axe—the divine feminine, the queerness—has survived to the present day. There are multiple versions of the lesbian pride flag, but the most striking in my opinion features the purple two-bitted axe against a black triangle on a purple field. But the flag is not widely recognized outside of lesbian circles, and when I see it used online there's sometimes an unfortunate whiff of transphobia—ironically at odds with the gender-fluid dynamic of Omphale and Herakles.

There was another powerful wife with an axe in the Greek canon about which Herodotus wrote. Aeschylus wrote tragedies, not histories; he could take as much license as he wanted, and he chose the axe as a symbol of women's revenge in his masterpiece, *The Oresteia*. Agamemnon was another king, brother to Menelaus, whose wife, Helen, was abducted by Paris, launching the Trojan War. Agamemnon sacrificed Iphigenia, his daughter with Clytemnestra, to obtain a favorable wind so he could sail off to war. When he came home,

Clytemnestra and her lover killed her husband and his lover in revenge.

Clytemnestra's weapon is an especially controversial one: scholars are still debating whether she used a sword or an axe to murder Agamemnon. The axe offers the symbolic load of sacrifice and slaughter, emphasizing the savagery of Clytemnestra in contrast to the civilized justice built to answer and punish her rage. Agamemnon was hit in the head, notes Malcolm Davies, the classics scholar who, in 1987, preceded me on Team Axe. This suggests she used something a little heavier than a sword.[14] And the sword is rarely attested during the Trojan War, some argue, but it definitely was by the time of Aeschylus.

These characters are most real not in histories but in poems and plays; whatever the weapon was, it was chosen for its dramatic effect. When Sophocles retold the story a few decades later in *Electra*, Clytemnestra "split his head with murderous axe, just as woodmen chop an oak."[15] If this axe is used to chop trees, it must be hafted. And this axe too is πέλεκυς, providing further evidence that Stesagoras's assassin also used a hand axe.

The violence that brought Stesagoras's life to an end was not fable but fact. There is no major lesson to be learned from his assassination, besides maybe that Croesus's power had limits. But the πέλεκυς is what has made the story worth telling for thousands of years.

Axes in ancient Greece were visible symbols of force. But in this case the power of the axe was concealed. A hand axe, or hatchet, is a weapon for the underestimated and ignored, whether Persian discontents or abandoned royal women. The axe was small enough and common enough not to register as an actual threat, but it was as sharp as the blade in any fasces.

The assassin chose a weapon whose power is generative and destructive, ubiquitous because it can do so many necessary and awful things. In the end, I do think the weapon used in Stesagoras's assassination was a hand axe or hatchet, a small axe to be concealed. They are the most portable, the easiest to pick up and carry. They are easy to grab when simple conflicts sprout into complicated violence, one or two sharp and practical edges equal to the brutal task of revenge.

Iron Shipbuilding Axe

1000 CE, L'Anse aux Meadows

Vikings weren't as into axes as you think.

Of course, they had tons of them, valued them, hoarded them. They certainly fought with them often enough. The necessity of the axe in daily life meant that more yeomen knew how to use an axe than a sword, so it was a tool that was often pressed into war with the average guys who made up the bulk of Viking raiding forces. And the Vikings used their own battleaxes—flared Dane axes (a higher-status Viking axe favored by wealthier warriors) and francisca throwing axes (the emblematic weapon of the Franks of medieval Western Europe).

But despite the fact that we might consider axes a classic Viking symbol, they didn't have much more prestige than a hair comb. Unless an axe was made of precious metal and stunningly enameled like the axe in Ahhotep's tomb, they weren't objects of great honor and pride like a fine sword. Instead, the axe was everyday, everywhere, part of making everything.

Shipbuilding was one of the most vital Viking occupations for the axe. Seafaring was central to the Viking identity. Vikings were widely traveled and spent years on campaigns of war and conquest and resource collection. Ships require the axe's sharpest and best work, the most highly skilled woodsmen, the best treated lumber fashioned extremely carefully. A log cabin with gaps in the walls can be plastered over or otherwise plugged against the elements, but that will not hold water when it comes to boats. The wear of constant use at sea and the curved planks necessary for shipbuilding made the axe a primary tool. Many axes had long blades and beards—the lower and sometimes upper edge of the blade extending down in a sort of curving stem. These forms allowed the axe to flatten the timber by bouncing off the wood in a pattern called "sprett-teljging"[1] or dig a boat out of a single piece of lumber.

But while they were off with axes, they needed someone at home to keep the hearth burning. The bulk of the labor of sustenance usually fell to women. The axe is, more than anything else, a tool to carve survival, shelter, food from the land, and there are few places where that work was more demanding than Viking-era Greenland.

FIVE

FREYDIS, WOMAN OF THE FOREST

When Freydis Eriksdottir killed five women in the woods of Northern Canada, she didn't have a very good reason. Her violence was petty, greedy, an afterthought. Yet it was as rooted in her family history as any blood feud, passed down from father to son to daughter. She slayed to show that she was as full a participant in the violence of the wilderness as any man, and that she deserved the same respect accorded to her bloodthirsty male relatives. Her privileged position as the daughter to Erik the Red and sister to Leif Erikson brought her the ego to enact the fantasies of a terrible temper—but unlike her father, she could not get away with it.

Iceland in the Viking era was tribal and isolated, but they did have a system to deal with people who killed their neighbors. Viking law in Iceland specified three degrees of punishment in response to killing. In the third degree, you were confined to certain areas for a period of time. In the first—the full outlawry—you were declared a "'man of the forest,' compelled to live apart."[2] Your property was confiscated, your friends were forbidden from sheltering you or helping you leave the country, and if you did manage to escape, you could never return.

It didn't take too long for Erik to merit the full outlawry.

Freydis's father was brought to an undesirable part of Iceland as a child because his dad killed someone and got them kicked out of Norway, and when Erik grew up he had basically the same conflict-management strategy. He killed two neighbors in a dispute over some damage from a landslide, and when that got him banished to another part of Iceland, he promptly killed another neighbor over an unreturned pillar. He managed to escape Iceland with his life and some compatriots, but Greenland is a harsh place to be in exile. He was a man of the forest in a land without trees.

In Greenland, Erik's crew found the ruins of a house and a boat in the southwest corner of the island and spent three years exploring, collecting goods, and hunting. When Erik came back to Iceland, he was not killed on sight. To the contrary: he talked a ton of people into coming back with him. The furs and hides and tusks he carried with him were a convincing advertisement for the riches in store for anyone who wanted to tag along—underlined by a clever bit of branding in calling the world's largest block of ice "Greenland."

It helped that the coast of Iceland was getting about as crowded as a barrel of fish. There was no good land left. Iceland was almost completely deforested within a hundred years of Nordic settlement. There weren't that many trees to begin with, because it's so far north, and the Icelanders needed trees for so many things: buildings, firewood, and ships. Because of the deforestation, raw lumber was often imported, still to be stripped down, planed, and shaped. Greenland was closer to Canada, where there were many trees.

Twenty-five ships left Iceland with Erik in 986 CE, each stuffed with supplies, cattle, and human cargo—both family and thralls, enslaved people who would do the real work of settling the frozen land. Only fourteen ships made it to Greenland. Some of them may have just turned around and gone back to Iceland, but others were trapped in a terrible and unique oceanic disaster. The Sea Hedge was a sort of Arctic Bermuda Triangle traversed by the "hafgerdingar," a wave "higher than the lofty mountain and resembling deep overhanging cliffs," according to a Norse saga. This Sea Hedge could have been caused by an "undersea earthquake"[3] or it could have simply been an optical illusion, a mirage in the middle of a terrible storm.

The Greenlanders who made it over didn't exactly know how to deal with going from tundra to ice sheet. Inuit and Dorset communities had lived there for thousands of years, moving from place to place as the climate and season demanded; when the Norse arrived, they were on the opposite end of the island. The immigrants tried to grow the staple pastoral crops that they had grown in Iceland, but it was mostly too cold for that. They eventually copied the Dorset and started hunting—caribou, walrus, hare, seals, and narwhal, whose tusks they took back to Europe to sell as unicorn horns.

The realities of their bare and scrubby land necessitated many trips back. The Vikings were quite a well-traveled people, cosmopolitan in a way that much of the rest of the world could not match. You might think communities at these small scales situated on an island would be insular, but historian Elizabeth Ashman Rowe told me that early Vikings made voyages as far as Baghdad and Russia. Erik's people weren't going on such grand adventures, however; they were making trips to import basic supplies from Europe, like iron and wood for their axes.

Erik's son Leif was converted to Christianity by the Norse king Olaf Tryggvason on one of these trips to Norway in the late tenth century, and when he got back to Greenland he set upon converting his family and the rest of the small community, only a few thousand strong. Leif was a charismatic leader, and most people were an easy sell—except Erik the Red, who was attached to his pagan religious traditions. Erik's wife, Thjodhild, became rather evangelical, though, withholding affection from her heathen husband and building a church for the other converts, "of whom there were many."[4]

The zeal of Erik's converted sons was such that one of them—Thorstein—became the subject of a resurrection story. Thorstein's neighbor Sigrid had a vision of Thorstein holding a whip before a horde of undead. Then Sigrid took ill and died, but she did not stay dead. She rose up and left her home to go to the neighbor's house, where she tried to get into bed with Thorstein.

Sigrid's husband tried to end this supernatural-messianic-erotic-thriller story by driving an axe through his wife's breast, but not before she infected Thorstein. He died and rose in turn with a message for his wife, Gudrid: she must change the local burial practices so that he would be mourned in a church and buried in consecrated earth, and in recompense she will remarry well and travel the world. She becomes the protagonist of her own saga, a hero of the great Nordic poetic history of Greenland.

The Vinland sagas were composed in the thirteenth century, hundreds of years after the lives of the first family of Greenland and around the time the Viking colony in Greenland finally petered out. Historians have debated whether the Saga of Erik the Red or the Saga of the Greenlanders was composed first and which one is more reliable, but they're both relatively realistic representations of the adventures and trials of the first European settlers of North America, give or take a zombie. The Thorstein anecdote is a work of art meant to entertain, and also part of a larger narrative that aims to keep alive the history of a society that did not keep a lot of records. It's not literal, but it's not a fantasy. More Herodotus than *The Shining*.

Freydis—and to a lesser extent her axe murders—were probably real. There's no archaeological evidence for Freydis, and she doesn't show up in the most reliable versions of Viking events, like the Anglo-Saxon Chronicle, a history compiled year by year beginning in the ninth century and distributed to a number of monasteries that updated their own versions through the twelfth century. But none of the researchers I talked to thought Freydis or her murders were simply myth.

There are two major stories in the sagas about Freydis. Neither feature zombies or gods or vampires, only human foibles and flaws: one is a story of bravery and colonialism and swords; the other, of greed and sports rivalries and axes.

We know almost as soon as we meet her in the Saga of the Greenlanders that Freydis is a menace. At first she just comes off as a brat; though she was illegitimate, she was an accepted daughter of Erik the Red, and so was basically a member of the first family of Greenland. Her status only increased with the success of half-brother Leif Erikson's "profitable and honorable"[5] expeditions to Vinland—the coast of Canada, specifically Newfoundland. There is a fairly well-documented Norse base camp in the area, L'Anse aux Meadows, where as many as seventy people bunked on short-term expeditions. Freydis wanted some of that for herself.

She approached two Norwegian brothers, Helgi and Finnbogi, to join her on a voyage because they had suitable boats. Each party agreed to bring thirty men and five women, but Freydis went ahead and brought an extra five men.

The daughter of Erik the Red is used to getting her way, the brothers probably thought. They'd put money in already, too late to back out now.

Helgi and Finnbogi probably knew that they had made a mistake as soon as Freydis landed in Vinland. The Icelandic brothers got there first and set up in one of Leif Erikson's houses. When Freydis arrived, she took issue with that. "'To me lent Leif the houses,' quoth she, 'and not to you.'"[6]

They replied, resentfully, "In malice are we brothers easily excelled by thee."[7] But they had little choice except to work with her. So the brothers built a new house by the shore of the lake and moved in.

Sports offered an opportunity to both work out and inflame tensions. Freydis and the brothers found a temporary camaraderie as winter set in. One of the games they probably played was with a ball and a bat, more like curling than baseball, and more violent too—the kind of amusement where "people definitely got hurt or killed."[8] There was also the popular autumn pastime of horse fighting, which is exactly the kind of animal-rights nightmare that it sounds like: "two stallions [urged on] by hitting them sticks [so] they would bite and paw each other."[9] That would have been a terrible use of resources while on an expedition far from home, but I don't know if any of these people had the impulse control to resist.

There could have been less violent delights as well: storytelling, reciting of poems and myths, of course, but also sick burns. "Another form of entertainment was a comparison," said Dr. Rowe, who is a professor of Viking history at Cambridge University. "Two men would take turns extolling a certain person—like, 'which of these two kings was better?' It's an exercise in partly insulting and partly finding reasons to praise your man or whoever it is you're talking about above the other person."

Between the horse fighting and the diss tracks, it's hardly surprising that the good vibes didn't last: "Evil reports and discord sprung up amongst them, and there was an end of the sports, and nobody came from the one house to the other, and so it went on for a long time during the winter."[10]

One morning, Freydis put on her husband's cloak and walked barefoot into the dew, headed straight for the brothers' house. Someone had already gotten up and left the door half open, so she stood at the threshold until Finnbogi noticed her. He asked her what she was doing there, and she asked him to have a heart-to-heart with her on a fallen tree outside the house.

"How art thou satisfied here?" she asked him, like she hadn't been pretty pissed at him for most of their expedition. Walking on eggshells, he answered that he loved "the land's fruitfulness"[11] but hated the bad blood between them, especially since he didn't know why it happened. Freydis used the leverage of the moment to ask if she could trade her ship for Helgi's bigger ship. Finnbogi had little choice but to say okay.

Back in Leif's house, Freydis woke up her husband, Thorvald, with her cold feet. He asked why she was cold and wet. She told him that she had gone to make an offer on Helgi's big boat, but they took it the wrong way and beat her soundly. Assuming this is a lie—Freydis gets little benefit of the doubt—it's worth wondering why she'd say it. She was getting everything she wanted. But perhaps what she really wanted all along was violence.

Or perhaps the guilt lever was too hard to resist. She framed Helgi's fake assault as Thorvald's fault: "But thou!

miserable man! wilt surely, neither avenge my disgrace nor thine own."[12] She followed this up with an ultimatum: avenge her fake honor, or divorce.

Once Freydis challenged her husband's manliness, he and his men had little choice but to murder. Accusing your husband of "unmanliness" and threatening divorce was a standard way to get him to be violent on your behalf. By the age of Erik the Red, Norse society was heavily militarized. Blood feuds were common. Violence was an expected part of masculinity, and there was little more important than masculinity. Freydis had enough status to let everyone know if Thorvald failed to defend her, which would have had real consequences—not least of which would be the loss of her wealth. Besides, the men on the expedition were the kind who were willing to kill.

Thorvald woke his men and told them to get their swords and axes and daggers. They snuck to the houses of the sleeping brothers, binding them before they could wake. Thorvald's men walked each of the thirty men out, one by one, and Freydis sentenced each of them to death in turn. But when all the men were dead, Freydis was still unsatisfied: "Only women remained, and them would no one kill."[13]

No man, at least. Even Viking violence has boundaries.

When Freydis learned that the men were done killing for the day, she was not deterred. "Give me an axe!" she said. They wouldn't kill the women themselves, but they also were not going to stop her. She "did not stop until they were all dead."[14]

In the afterglow of her grim accomplishments, she told

the thirty-five men she brought that if they made it back to Greenland, she would "have anyone who tells of these events killed. We will say that they remained behind here when we took our leave."[15] She had no attack of conscience later. The Saga of the Greenlanders records that "they went back to their house after this evil work, and Freydis did not appear otherwise than as if she had done well."[16] Perhaps she thought she had done well. Maybe she was even pleased with how the day unfolded, like she'd been planning it since before she darkened Finnbogi's door to see if he wanted to make some money.

She returned with her ships loaded with goods, one of the most valuable cargos ever seen in the Greenland bay. Everything went fine at first—Freydis's farm had prospered in her absence, and she paid out her men handsomely, hoping to enlist them as conspirators. She stuck to her farm, avoiding the glory of her profitable and honorable haul. Perhaps people would simply forget about the nearly forty men and women who didn't return and accept it as another harsh consequence of a rough land.

Mass murders never stay secret. Especially in a place like Greenland, only a few thousand strong; it was basically a small town. Before long, Leif had heard about it; he did a little *Law & Order: Greenland* by torturing three of the men to find out what had happened. Their stories "agreed in every detail."[17]

"I am not the one to deal my sister, Freydis, the punishment she deserves," said Leif, who, it should be pointed out again, had just tortured a bunch of people. "But I predict that [her] descendants should not get along well in the world."[18]

There would be no more profitable adventures in North America. Once the truth was out and her brother deserted her, her reputation was gone; "no one expected anything but evil from [her]."[19]

Freydis's reputation wasn't completely ruined for all time. The Saga of Erik the Red—the one with the zombies—features a story about her that makes her sound brave and heroic, a protector of Greenlanders rather than someone who would murder her own.

One winter in Canada, the Greenlanders were trading with the native inhabitants, probably from Beohuck or Mi'kmaq tribes. Suddenly a bull belonging to the Norse emerged from the woods and scared them off. Some weeks later, these warriors came back and began thoroughly kicking Viking ass. One projectile, "about the size of a sheep's gut and black in color,"[20] made a very threatening sound when it landed, and the men all fled.

Then Freydis came out of her tent to see what was going on. She was heavily pregnant, and in no mood for any of this. Her first move was to goad the men for their lack of manliness, but the men ignored her; perhaps she used this particular tactic too often. They left the expectant mother to fend for herself; she tried to keep up but could not.

As she tried to hurry, she came upon the body of a man

killed with a stone, his sword lying beside his body. Freydis turned around to face the Indigenous warriors chasing her. Taking the sword from the dead man, she turned and exposed her breasts. She "smacked" the broadside of the sword against her chest, and the men retreated to their boats to row away in fear and confusion, allowing for the men in her party to enjoy a later triumph. There the story ends, and Freydis does not come up again.

The arresting image of a woman taking her nakedness and the heft of her pregnancy and turning it into something to be feared is deeply rooted in multiple religious traditions. In his book *The Bare Sarked Warrior*, Cornell University professor Oren Falk places the visual impact of bare-breasted Freydis into European medieval tradition, connecting it to bare-breasted motifs from Christian and American revolutionary art. Her zeal to colonize North America reminds me of the bare-breasted Amazon in the state flag and seal of Virginia. Still, this bare-sarked (or, "bare-breasted") Freydis presents a stark moral contrast to the boat thief and axe murderer Freydis of the other saga.

And so, perhaps, the sword is a fitting contrast to her axe: where there's honor, there's a sword, not an axe.

"The sword is like a Mercedes among weapons," said Dr. Leszek Gardela, researcher at the National Museum of Denmark. The metal of a sword can only be put to tasks of violence—and there's a bigger learning curve for using it well. It was a privilege to have and to wield, the most "elite" weapon. Swords were powerful and sophisticated, the kind of thing Norse gods held. Almost none of the pagan gods wielded axes; perhaps it was too human an instrument, too applicable to our dirty tasks.

The axe was a low, domestic implement, used not just for the exciting violence that animated so much of Viking life but also for the drudgery of trying to survive. For a long time, researchers assumed that any Viking skeleton found with an axe or a sword or a dagger was a man, and that women's graves were marked by brooches and other ornaments. But that wasn't true at all. Many Viking women were buried with axes. Some of them were warriors. But others were simply engaged in the daily work of life in cold and sometimes inhospitable lands. The domestic axe, an everyday implement with immense value, was as powerful as the battleaxe. It was a tool of survival, and there's nothing more powerful than survival.

Survival comes with its own kind of magic. Dr. Gardela told me about a Nordic tradition practiced into the early twentieth century in which a person with a leg or foot ailment placed their limb upon a tree stump while an axe was slowly brought down around it, to symbolize amputation. This was used not just for physical ailments but the concept of bad luck. Where the sword's ability to piece and lacerate is powerful, the axe's power lies in its ability to chop and cleave one thing from another. "The axe is a tool that defines things," said Dr. Gardela.

The Vikings defined themselves not so much as savage colonizers than as fearless travelers. That occupation depended on their ability to build and launch huge ships, and each part of the process demanded not swords but axes.

Freydis's ships were launched just to find more of the trees they needed to make more ships. Great trees then had to be split into many long planks and curved to form a ship's keel and hull. Once the planks were cut down, they had to be whacked into shape with a long-handled hatchet and shaped with a bearded axe. The axe had many parts to play in the life cycle of the ship.

In Freydis's story, that extended to the conflict over the ship. Freydis asked for an axe because she didn't need to intimidate her victims as she did her attackers. The axe underscores the domesticity of that violence—and how her violence comes into conflict with what we expect from women.

These stories of survival and domestic life and adventure include a whiff of the soap-operatic sexual intrigue that has powered popular fiction since before Clytemnestra. Erik's wife refusing him in the name of Christ. Poor undead Sigrid trying to climb into bed with another woman's husband. Freydis's breasts, bare and struck by the steel phallus of the sword. The very presence of those five women with the brothers' thirty men. Freydis slipping out of her own home and into another man's, barefoot, silent, provocative. Her wet feet waking her husband. The text's description of her fabricated ill-treatment at the hands of her business partners did not imply sexual violence, but part of the basic tension in the Vinland sagas is sexual: conflicts are caused by too many men and not enough women.

Freydis was in a unique position. Her violence did not come from a lack of power. She was under the protection of her father and brother and held power on her own; she was the leader of the trip, not just financially but through the force of her personality. A woman alone in the forest with a man has an inherent vulnerability, but Freydis seemed not to feel it, on that day or any other. Yet she was willing to invoke that vulnerability to goad her husband, a weapon she held as easily as axe and sword.

She is the catalyst for events, but as Dr. Falk wrote, "the actual triumph belongs to those of the correct sex."[21] In her bare-sarked saga, she is permitted to be the exception because she faces an enemy that is also an exception, a despised out-group who nonetheless are capable of soundly defeating the fearsome Vikings and proving to be the one society they cannot raid and conquer. The story is not so much evidence of an individual woman's bravery as an eroticized distraction. Her transgression is so intense that it becomes salvation. At least when it comes to the sword.

As Falk points out, the Greenlanders' saga, by leaving her lost in Newfoundland forever, makes her into a woman of the forest, a murdering outlaw.[22] This outcome could not have been a surprise to anyone who knew Freydis. "She's very much her father's daughter, [with] a kind of bloodthirsty, aggressive tendency," said Dr. Rowe. "And so just as her father was exiled twice for killing people, she apparently has that same kind of personality."

Like her father, she was frozen out of the land of her birth; unlike her father, there was nowhere she could go to redeem herself. There was no great haul of furs and timber to sway her small world of Greenland back on her side. Neither axe nor sword had the power to redeem her reputation.

Executioner's Axe

circa 1510, London, England

The axe is a timeless choice for the status-conscious condemned. For the innumerable aristocrats sentenced to death in the Tudor era, a sharp axe used by a practiced executioner was a stately way to die, though not as stylish as a sword. And there were reasons aside from fashion to wish for the axe. After all, decapitation is less painful and humiliating than hanging, burning alive, or being torn apart.

Henry VIII unconsciously echoed the Shang approach to the axe as the ultimate display of his power. In the Tower of London hung a "roughly made" "heading axe,"[1] its head almost obscenely long (sixteen inches). There is a resemblance to the Dane axe, with its broad edge splaying from a thin butt. Its blade curves down, with a span wide enough to cover most necks—ten inches. The axe on display was "neither sharp nor polished"[2] and looked as if it could crush better than it could cut.

In a highly stratified court, a person's status could save them from being drawn and quartered—but it could not always buy them a sharp axe and a steady hand. The doomed men and women knew that the man holding the axe was responsible for the accuracy and sharpness of the axe, and they were doing a kindness to the person whom they were beheading by doing their work with as much brutal efficiency as they could muster.

SIX

PIGMEN, GARGOYLES, BLUNDERING YOUTHS

Reputation is everything in a small community, whether it's a few thousand farmers trying to survive a harsh and isolated life on the coast of Greenland or a royal court enjoying the finest luxuries that Tudor England could provide. At the Tudor court, rumors and gossip ruled the lives of the richest and noblest people. When that gossip curdled, there were no gifts, no furs, no homes, no level of piety good enough to protect against King Henry VIII's system of justice. Members of the nobility and the wider Tudor court considered the class implications of every detail in their lives from their hood to their bowel movements. When they had the sorry opportunity to figure out the stylistic details of their own deaths, they weren't just thinking about the pain and suffering that could be lessened by being beheaded instead of hung. They were thinking about decorum, legacy, piety. The axe had a time-honored dignity that brought order and meaning to an execution.

Henry VIII stands out in history not just for the Sea Hedge of executions he ordered but for how many of them spill from his personal animus. Execution was a huge part of early modern justice, and from early in his reign Henry used it unjustly on targets close at hand in the highest social classes. And when he wanted to inflict additional cruelty, he would strategically withhold the strange mercy of the executioner and his axe.

Here we have seven of the highest-status people so delivered to death: Edmund Dudley, Richard Empson, Edward Stafford, Thomases More and Cromwell, Catherine Howard, and Margaret Pole (née Plantagenet). Their heads, removed cleanly or raggedly, form the arc of the king's abuse of execution for personal grudges that grew more volatile and illogical as his terrible reign wore on and his exercise of state violence escalated from a cold, cruel, considered ritual to the rough-edged violence of the axe murderer.

Richard Empson and Edmund Dudley were middle-class boys pursuing nobility by doing King Henry VII's—Henry VIII's father's—dirty legal work. Empson was the son of "a minor property owner [who was] rich enough to pay for his son's legal education, but not much more."[3] Dudley, the son of a baron, was sharp-witted and memorable in the courtroom, an able advocate for expanding kingly rights, and a character ready for *Law & Order: Tudor Court*. Dudley's charisma enabled his rapid rise—a trait he passed down to his grandson Robert, who in a couple generations would ruin Elizabeth I romantically.

Empson was no overnight sensation; his was a long and slow rise through the York ranks, where he was in charge of finding new revenue sources for the Crown, aggressively pursuing debts, customs fees, rents, fines, licenses, taxes—all the annoying things no one wants to spend their money on. Empson and Dudley were very effective in enriching Henry VII's royal coffers and raising their own status, though they also became excellent targets for an angry public. "Thou pigman, thou gargoyle," one 1504 poem castigated Empson."[4]

Henry VII had been king for twenty years; his reign was never quite comfortable, but England was much more stable than during the War of the Roses. In the last few years before his 1509 death, as Henry VII grew paranoid and out of touch, Empson and Dudley forged a profitable and strategic alliance. Some of the debts that Empson and Dudley collected were not actually meant to be paid—they were threats, ways to make gentry do what Henry VII wanted (perhaps a bit like student loans). Dudley, especially, was good at using his knowledge to give himself real estate advantages.

They weren't the only middle managers making themselves rich in the dotage of Henry VII. But while other acolytes of Henry VII's aggressive fundraising were more or less seen as men of their word, Empson and Dudley were both seen as "archetypal low-born men on the make" and "smooth talking deceivers."[5] Dudley's charisma, Empson's extravagances, and both men's greed put targets on their backs basically the moment Henry VII finally expired. Only three days after the old king's death, they were arrested.

The primary charge against these administrators was, bizarrely, that they had tried to influence the order of succession to their own advantage. Henry VIII was only eighteen years old, soon to marry a twenty-three-year-old Catherine of Aragon. There was no reason whatsoever to worry about the fecundity of his union; he had decades to produce an heir. And yet it was already an existential struggle for him, a preview of a life obsessed with furthering his dynasty. It was as if he knew already that his reign would be cursed by this very issue, that he would again and again experience miscarriage, infant death, and impotence throughout his many terrible marriages.

Whatever their pretext, the arrests had the effect that they needed to: the two high-profile secretaries of the previous king were blamed for all of his extortions, and his son, the new king, could position himself with a clean slate, above suspicion for how he paid for all of his crowns and gowns and jousts. And because the charge against Dudley and Empson had to do not with bookkeeping but with succession, it became a matter of treason. Worse than the shame of their fall from grace, perhaps even worse than the death sentence, was the method to which they were condemned.

Drawing and quartering was reserved for treason, a human-sacrifice ritual hard to exceed in its degradation and befouling of the victim's body. It was a process utterly removed from justice, more like a snuff film performed live for an audience. The punishment was devised in the mid-thirteenth century for noble traitors and inconvenient princes. It was intense, too "indecent"[6] to do to a woman—female traitors were burned at the stake instead.

First the condemned was "tortured and starved."[7] On the day of, they were dragged by a horse to the site (in the Tudor era, Tyburn Hill) by a horse, sometimes tied to a plank. The condemned man was first hung using the short-drop method. The long-drop method is the modern(ish) method of hanging—what happens on a scaffold, from which the body falls farther and the neck breaks quickly. In the short-drop method, the body does not drop as far, and death results from the body's struggle, which tightens the noose and leads to strangulation. But in drawing and quartering, they didn't finish the process, so that the condemned would be alive while they cut off his genitals and threw them into a fire. Sometimes they finished the job there by opening up his stomach and throwing his entrails into the same fire. Then the heart. Then the body chopped into quarters. Sometimes the quartering process was performed by horses, each limb tied to a different beast headed in a different direction. Then, finally, the victim was beheaded, and the dismembered parts of their body presented to the king. If the crimes for which he was executed were particularly heinous, all five mutilated pieces of his body would go on the castle wall or London Bridge as a warning.

You can see why people did not want that to happen to them. "Many who were found guilty of treason begged the monarch to grant the more dignified death of the block," wrote Toba Malka Friedman in her 2009 dissertation on Tudor execution of the nobility.[8] It was often the most principled objectors who refused to perform the necessary contrition of begging. In 1535, the king had six hermits who refused to recognize him as the head of the church drawn and quartered; nothing angered him as much as noncompliance, and so those who put God before him received the cruelest ends.

Dudley and Empson had a year to consider their short and grim future in jail. Dudley spent his time writing a book about monarchy and society called *The Tree of Commonwealth* that is a valuable insight in the social stratifications of the day as well as flattering to Henry VIII's idea of what a king should be. It wasn't enough to save them from death, but at least they avoided being drawn and quartered. Instead, they got an old-fashioned English sort of death: a beheading at Tower Hill with a great big axe.

Henry VIII came to the throne on a wave of popularity and good feeling, and he focused on forgiving his father's fines and handing out favors and titles freely. His good looks, charisma, and intelligence made everyone in the country want to give him all the benefit of the doubt they could give. The execution of Empson and Dudley was seen as perhaps excessive but in aid of good governmental policies like not overtaxing.

As his reign progressed and the heads began to pile up, there was an increasing awareness among the nobility and the common folk that his enthusiasm for executing people who weren't even rivals but simply not yes-men was no exercise of strength. He inherited his father's tendency toward suspicion and conspiracy, especially when it came to anyone who rivaled him in glamour, popularity, or nobility. Not just administrators but members of the nobility became targets for elimination under paranoid Tudor rule.

Edward Stafford, the Duke of Buckingham and Henry's first cousin, was able to play along and pretend he wasn't a rival for the first decade of the young king's reign. They had a lot of family history that should have kept him safe—Stafford's father was beheaded for supporting the former King Henry VII in the War of the Roses. But eventually Stafford committed the treason of acknowledging his own power and found himself following his father to the chopping block.

The young duke—the only duke in the country at the time—was no major player in foreign or domestic policy. But he had no shortage of glamour, wearing gowns worth millions of dollars to state weddings. Despite the violation of dress code etiquette, he kept up a friendly relationship with the king and court, hosting Henry to "excellent cheer" as late as 1519.[9] Stafford was happy to suck up to the king to a certain extent, but he had his own life and was not focused on attaining the "frequent, personal access" necessary to staying in the king's favor.[10]

Part of the 1510s Henry charm offensive was based on repairing relations with France as opulently as possible. Edward Stafford was against this from the start, setting him against Cardinal Wolsey, whose personality was as similarly flair-laden as the king's. Though Stafford participated in the pageantry of Henry VIII's meeting with French king Francis I at the Field of Cloth and Gold in 1520, he was against any concessions to a French alliance—pitting him unluckily against the king. Stafford left court and hid at his home in Thornbury, hoping that Henry would forget about him.

But he didn't. The more distance Stafford tried to create, the more fixated the king became on his threat to the throne, collecting and sifting through all available gossip. Stafford participated in this a little too, allegedly. He's unusual among Henry's victims in that there was actual evidence (mostly secondhand from servants but plentiful nonetheless) that he was quite guilty of saying unflattering things about the king.

In April 1521, Stafford was summoned to the king's residence at Greenwich. He had no idea he was in serious trouble until he realized he was being followed. He went straight to his frenemy Cardinal Wolsey's wine cellar and had a drink before getting arrested and led to the Tower of London. At his hasty trial in Winchester, the king personally examined witnesses. The jury had little choice but to find him guilty, but they could not deliver the verdict to his face; Stafford had to exhort the jury personally to face up to it and condemn him to his death.

On the barge journey back to the Tower of London, the axe was turned toward Stafford to indicate that he awaited the same fate as his father. The Duke of Buckingham "died in agony, beheaded on 17 May by a bungling executioner who took three strokes of the axe to sever his head."[11]

The duke surely served as an early warning to the rest of the nobility: it could happen to them. Buckingham had royal blood, standing, connections, wealth, glamour, good cheer— everything every rich person thought they needed to live safely if not happily under the constant pressure of Henry

VIII's convivial but jealous eye. But none of this protected him from an early and painful death.

In Tudor-era depictions of the Tower of London, Tower Hill is often off in the distance with a scaffold—on the perimeter but still highly visible. And because the bloodstained ground was outside of tower grounds, it was an accessible space, open to all. Indeed, that was part of the charm. Executions were a regular feature in British entertainment, even when other large public gatherings were discouraged for fear of rioting, because they reminded citizens of the power of the state.

And it was a thrill. Death was a constant companion in plague-soaked England. Death and the art of dying well were a constant preoccupation, and execution offered Londoners a vicarious and cathartic way to confront their fear. They could judge how others died and remind themselves that it really could be worse. At well-attended executions, there was a crowd participation element—at a 1596 execution of a Jewish doctor, his proclamation of his love of Christ drew gruesome laughter. Postexecution coverage and commentary could be deeply detailed, down to "the length of time bodies burned [or] the number of axe strokes."[12] And the lessons did not end with death: after execution, heads and limbs were placed on pikes at the gates of London, a gruesome, rotting show of power.

Tower Hill was an easy dead man's walk from the most important of all Henry VIII's many estates. The Tower of London was like Buckingham Palace crossed with the Tyburn hanging tree, its fanciest, most glamorous lodgings;

hospitality; and entertainment grimly caked with the dust of thousands condemned. The castle has been a symbol of brutal grandeur since the eleventh-century Norman invasion, making it almost as ancient to Henry and his victims as they are to us, but it was in constant use. Henry prized the comfort and luxury of his personal surroundings, and he was constantly seeking novelty by renovating or acquiring new domiciles; in the tower, he "refurbished the Great Hall" and extended the private quarters and kitchens.[13]

A stay in the tower was a mark of a life lived successfully enough to personally piss off the monarch. The nobility of the Tower condemned and especially their submission to a decorous execution was an important part of the morality play, showing that even the nobility had consequences to face when they came into conflict with the king.

And there was reputational benefit to the executed. Being troublesome enough to get executed after a stay in the tower meant that you very much mattered. People in Tudor England believed strongly in the afterlife, and most were Catholic: wealth and status have serious bearing on the afterlife: "In the act of being defeated by the crown, a traitorous noble could, nevertheless, snatch victory for himself and, paradoxically, for his sovereign by reinserting himself within the ranks of elite society thereby justifying his own position while justifying that of the monarch's."[14]

It's kind of like a rock star dying at age twenty-seven: these departing nobles became part of an elite club forever. Would you rather be a forgotten courtier or in league with dukes and cardinals and queens? Friedman writes about an inverted pageant, in which the noble beheaded is a sort of carnival king: the undisputed center of attention, taking on all of the frustrations everyone has with the king that they can't say because they themselves would be drawn and quartered. Beheadings were a way to show one's character and even perform masculinity or femininity to one's last breath. No man could cry on the scaffold and still die a manly death.

Thomas More was sentenced to be drawn and quartered at Tyburn, but like Empson and Dudley, his sentence was commuted to beheading at Tower Hill. More—a firm supporter of Catherine of Aragon, the queen whom Henry tried to dispose of because he had tired of her—was taken into the Crown's custody in 1534, languishing for fourteen months in his room at the tower. As with other prisoners there, he had some of the pleasures of a wealthy home; he was allowed to write letters and receive visitors. At first he had tapestries for warmth and plenty of books, but the longer he stonewalled the king by refusing to acknowledge Anne Boleyn as queen, the worse his conditions became. Thomas was a proud man who, like the king, required a certain degree of finery. Eventually Henry agreed not to have him drawn and quartered, and allowed his longtime advisor to have a few worldly comforts—in his last weeks, an Italian merchant friend sent More wine and meat and a fine silk suit in which to die. And More's daughter and wife were allowed to bury him, as long as he promised not to give a barn burner of a speech in his final moments.

The final earthly comforts, the assurance of his burial, and the beheading allowed Thomas More to keep his mood light in the last thirty minutes of his life on July 6, 1535. "Is there really any difference between dying, one's head on a block under the axe, and dying, pining in bed?" biographer Daniel Sargeant wrote in 1933.[15] More was confronted to the end with administrative tasks, telling a woman who demanded some papers of him, "Good woman, have Patience but for an Hour and the King will rid me of the Care I have for those Papers, and every thing else."[16] He met a distressed, suicidal man on the way to the gallows, who came to him in desperation, and More told the man he would pray for him if he returned the favor. As he climbed the rickety ladder to his death, he asked for help from a lieutenant—but assured the officer that he would be able to get down himself.

More kissed his executioner and blindfolded himself. He said: "pluck up thy spirit, man, and be not afraid to do thy office; my neck is very short." More had a lot of forgiveness in him; he also didn't blame Henry, the man actually responsible for his impending headlessness. But he did not blame the man who held the axe for his death, closer to administrator than murderer in some ways. Kindness to the executioner was a part of the pageant of execution, as well as a good way for the condemned to limit their own pain with a sharp and accurate axe. "Take heed that thou strike not awry," as More put it.[17]

As he laid his short neck on the block, he had one last joke, one final vanity from a man who loved fine things: he swept his beard out of the way and asked his executioner to pardon his whiskers, as they were innocent of any treason. The blade was swift and true, needing only one stroke.

Though a beheading at Tower Hill confirmed their standing, Henry's targets could not go to the block as noblemen.

Execution is, of course, a human-sacrifice ritual. Nowhere was this more true than in Tudor England, a "world filled with intricate and symbolic markers."[18] Part of the inverted pageant of execution meant divesting one's nobility, returning to the status of a common man, even as their treatment evinced that they were not. There was a whole ceremony for this "shameful removal of rank"[19] dating back to medieval execution practices, in which the condemned was humiliated and dehumanized in the lead-up to their execution. The condemned man's crimes were read aloud, and then a herald would "take down the offender's crest, throwing it violently to the floor followed in like manner by his banner and sword [and] cast them into a ditch."[20] Having gotten that out of the way, their time on the block itself was "not meant to be a moment of degradation but rather an occasion for renewal."[21] The execution could be an elevation, a restoration to full status.

Thomas Cromwell spent his life accruing status, rising from a merchant's son to the Earl of Essex. Cromwell had been Henry's greatest ally during his quest to make Anne Boleyn queen, but after her beheading (which we'll discuss soon), he fell from the king's ever-more-capricious favor. Just two months after he was granted the earldom, he lost it. On June 10, 1540, Cromwell was arrested as the marriage he arranged between Henry and Anne of Cleves fell apart. After removing Cromwell's rank, the king sent out a special announcement that Cromwell was no longer his chief minister, lord great chamberlain, and Earl of Essex, but simply "Thomas Cromwell, Cloth Carder."

Yet despite the loss of hard-won honor, Cromwell faced

his end cheerfully. Perhaps it was a relief after the tension and mania of trying to accomplish Henry VIII's goals. He was in good spirits on his way to the gallows on July 28, 1540, according to one account. On his way across the street, he encountered another condemned noble, the baron Hungerford. Cromwell told him cheerfully that if they repented heartily, they would have a joyful dinner in the afterlife after breakfast with the axe.

Cromwell rose to his brief nobility because of his ability to get things done. And he knew the task ahead of him. Using wording echoed by many others, including Anne Boleyn, he said: "I am come hither to die and not to purge myself."[22] An administrator fixes problems, and he was the problem to be fixed that day.

Cromwell was world-class at accomplishing tasks bloody and dry. The executioner was not in his class. The headsman "bungled the job badly,"[23] and Cromwell's head went to a pike over the London Tower, raggedly cut at the neck.

Henry would never be restored to the adoration that met his accession, but he was not an entirely unpopular king. His kingdom "owed its successes and virtues to better and greater men about him,"[24] but he was a man who understood propaganda and how to take credit. The savagery of the executions were in full view, but bloodthirst is real and he sated it, which will always count for a lot with some people. And yet as with Vikings, there were some boundaries when it came to killing women.

Just as women were not drawn and quartered but burned at the stake, noblewomen did not suffer the public humiliation of walking across the street to be publicly killed. Only ten people were executed within the actual castle walls of the Tower of London, most of them women. These executions were not entirely private, but they were not for public view. A woman out of place was too powerful a sign of disorder to kill her out in front of everybody. By the time Henry was ready to execute Anne Boleyn, disorder was everywhere. Catherine of Aragon's death in January seemed like a boon to the struggling marriage, but by February Henry had suffered a concussion from a near-fatal jousting accident, and Anne had a miscarriage. The more he tried to suppress Catholicism, the harder the people fought back. His towering rage and paranoia were still somehow escalating.

Yet Anne kept her sense of the external order to the end. She had a decade's experience in making her dramatic mark at court, and this was her final pageant. She was perhaps the most glamorous of any Tudor victim, in ermine, a fur-trimmed patterned gray gown, a crimson petticoat, and her famous French hood. Her sense of style and status extended to the cutting edge.

The king would not commute her beheading, but he got Anne the best executioner's blade. She was spared the axe; instead, a swordsman from Calais cut her little neck.

To the end, Anne was as quick-witted as any of her husband's advisors. People are still debating whether her description of her husband as "merciful" and "a good, a gentle, and a sovereign lord" was supposed to be ironic "words of defiance" or just a part of the Tudor script of a good and penitent death.[25] That was classic Boleyn: even as she faced death with bravery, dignity, and decorum, she couldn't resist getting in a layered little dig.

Catherine Howard had no such pretenses.

Henry VIII was disordered in 1536, but by 1541 he was completely out of control. His reputation still held sway among his people at large, but the people who knew him feared and were disgusted by him. He was an impotent man with a gross wound on his leg that wouldn't heal. So he married a beautiful teenager, and when the completely predictable thing happened and she took an interest in a man her own age, he decided to swiftly decapitate her.

Like all of these victims, Catherine deserved another end, in a different book about struggle and loss and triumph. Catherine belongs in a teen drama, planning outrageously glamorous outfits and plotting secret hookups. She was a sweet girl, but definitely not a woman. She was not older than twenty-one when she was executed. Her life was so brief and we know such intermittent details, but she is clearly a teenager in my mind—a polite, polished, if slightly spoiled child, someone who needs time to mature and figure herself out. Someone with peaks and disappointments ahead of her that do not require her to practice flinging her hands out for the executioner's block.

The queen "descended into a blind panic" when she was informed that it was time to get on the barge to the Tower of London.[26] Dressed in black velvet and a French hood, she reluctantly boarded. On the way to the tower, she passed the heads of her alleged lovers, Francis Dereham and Thomas Culpeper.

Once she was settled into Anne's old room at the Tower, she was able to calm down and think about how she wanted the next morning to go. Catherine was a person who always wanted to follow the correct etiquette. She was very concerned about process and protocol, striving to carry out her rituals flawlessly. She also gave thought to her outfit, though she didn't have time to be the fashion plate that Anne was.

Nor did she insist upon the sword instead of the axe. She didn't have the standing to make many additional requests, and there was something else she wanted. From her cell, she asked the staff to bring her the executioner's block. Again and again, she practiced how prettily she could lay her lovely neck upon the block.

The next morning brought a typical February chill. Catherine had requested a private death, but the relative privacy of the tower walls was all she would get. She wore a dark dress. Her speech was not recorded except to say that it well expressed the expected emotions about the Lord, the Church, and the king. There is a persistent rumor that she said she would rather die as the wife of Culpeper, misinformation that confirmed a romantic narrative rather than corresponding to the etiquette of the day, as Catherine always sought to do. She wanted to be remembered as a perfect queen, not highlight her rumored infidelity.

It was a good death, someone told her brother a few weeks later. Catherine performed to expectations, and the executioner responded with a single swift stroke. The pageantry and rituals of death were of great importance to the condemned as they approached "the most solemn and important moment of their lives."[27] But not all women went gently, pretty, like a scrap of lace cut in two. Some women fought like hell.

Margaret Pole was one of the richest landowners in England in the early sixteenth century. Like Edward Stafford, she was royal enough to threaten Henry—a Plantagenet by birth, niece to two kings. When she was finally awarded her father's estate in the 1510s, she owned four different estates—castles, manors, and palaces, one of which had a moat. Women couldn't hold public office, but in private law they could have as much authority as anyone else, and so she managed the staff, inventory, books, and other major tasks as her husband would have done if he hadn't died. She had to engage in "conspicuous consumption served to overawe those under a noble's authority."[28] She was even expected to turn out resources and muster men for military obligations—much like her dear friend Catherine of Aragon, who rode out into battle while heavily pregnant in this same time period. Margaret was the Countess of Salisbury, the only woman in the country to have a peerage title by her own right of descent—until Anne Boleyn was made marquess before her coronation.

Margaret was very Catholic. The Poles and the Plantagenets were against the new teachings of Protestantism, but she wouldn't say anything to the king's face. Even as rich as Margaret was, she couldn't afford it. The atmosphere was not as utterly terrifying as it would be in the 1540s, but in 1533 everyone in royal circles was beginning to be quite reasonably worried after how thoroughly and furiously Henry turned on Catherine. Catherine was a Spanish princess, hard to execute without upsetting diplomatic relations, and a beloved queen with all the gravity and charisma and magic that came with the role. Before it was conceivable that a queen would ever be executed. Margaret certainly didn't approve of the break with Rome, but she kept quiet about it.

It was her loyalty to the queen and the princess, Mary, that doomed Margaret. Margaret was governess to the sole heir to the throne from 1520 to 1533. Obviously, this was a position of huge trust and prestige, and an enormous responsibility; Margaret "strove to support and help Mary through some of the most traumatic years of her life."[29] And when the king wanted to bring Mary back to court, Margaret was resistant to give the position up. Margaret promised to house Mary in the style of a princess out of her own pocket, but she had no luck in convincing the king to let her keep Mary. She was also evasive about returning certain of the princess's household goods—her jewels and fine plates—to the king when he was trying to starve out Catherine of Aragorn and her child so she would abdicate. And she made the unwise choice to not immediately give the king anything that he looked at longingly, like one of her lovely castles.

After Anne's execution, the king's paranoia was gaining momentum and intensity. He'd already killed so many—friends, relatives, lovers, wives, his most trusted advisors and loved ones. All the Poles the Crown could get their hands on were arrested, including Margaret and her sons Henry and Reginald. Henry was executed, Reginald released, and Margaret was held in the tower.

Where Henry kind of forgot about her for three years. Sometimes maniacs get distracted. Her confinement was strict, but he did not leave her in there cold and hungry. She had servants and fine clothing and food, and sometimes she got the chance to go outside. She was an old lady by this point; contemporary witnesses to her death thought that she was in her eighties, though she was not even seventy yet. Just two months before her death, the king spent a bunch of money on new clothes for her.

There was basically no forethought to her death. When the day came in April 1541, there was not even a proper scaffold. She was not told the month or the week or the night beforehand; she was told that morning, and she was hustled out to the Tower Green at 7 a.m., so it couldn't have been more than a couple hours before she died.

Just as Margaret did not accept Mary's reduced circumstances, she didn't accept her own death. She did not give a sober speech at the end of her life, sucking up to Henry VIII one last time. She did not have time to game it out, think about how the consequences might shake out for her sons and grandchildren. She wasn't thinking about the possibility of being burned alive. She didn't even know what she was accused of.

There was no time to have last conversations. There was no time to consider the appropriate and ladylike way to approach the scaffold. There was no time to plan out an outfit.

There was certainly no time to have the conversations even an imprisoned woman puts off, to make the necessary apologies and seek the forgiveness necessary for a good death. She was not given any time to talk to a religious official, to get her soul nice and clean and ready for God.

She approached the block with dignity and made a request for prayers for the royal family, especially Mary. But she lost it as it came time to put her neck on the block. At that moment, it became clear that she would not get the time that other members of the nobility get to prepare a few final words. She would not get to present herself to the 150 assembled people in that corner of the tower as a lady of high birth. Most devastating to a future Catholic martyr, she could not prepare her prayers and ready herself for heaven.

After so many years at the Tudor court, Margaret would have known what a usual execution for a member of the nobility looked like. This was thrown together at best. Usually a person climbed up on a scaffold and knelt before the block, with a pillow at their knees. There was no such nicety for the countess.

And the executioner. The axe.

The usual guy—a seasoned professional—"had been sent north to deal with the executions of certain of the northern rebels." The executioner was a trembling child: a "wretched and blundering youth."[30] Maybe he'd been the best woodsman in the king's retinue, but he was untrained to deal with the reality of ending a human life.

There is debate about what happened as Margaret approached her last moments. Certainly there is no description of Margaret flinging out her arms in consent in a regal gown. There were some accounts that had her refusing to set her head on the block, saying, "So should traitors do, and I am none." Once restrained, she refused to set her head still: "she bid him if he would have her head, to get it as he could: so that he was constrained to fetch it off slovenly."[31] Apparently, she struggled so much that her wig came up and revealed her gray hair.

With the opportunity for prayer comes the opportunity to speak one's piece. An intact body is a chance to make one final sartorial statement too; as we come to Henry's beheaded queens, we will hear about the care they took with their final outfit. But more important are the last words. In most cases, a person who is to be beheaded has the chance to say a few words and pray as they climb the scaffold and await the blade. "The individual who suffered the stroke of the axe had his posterity and his immortality to orchestrate as he faced human society for the final time," Friedman wrote. "The composure needed to die well by the axe marked an achievement of self-control and indicated the dignity of the beheaded."

But their greatest concern was God. Their last words were a prayer, a hope that they might find the mercy their king withheld on the other side. A body neatly cleaved and comprehensible, a holy relic worth burying—and not a headless, quartered horror.

Which is what Margaret became. Most accounts agree that the executioner botched the job. He whacked away at her frail body, making whatever pain she experienced excruciating and probably messy. The young man "literally hacked her head and shoulders to pieces."[32]

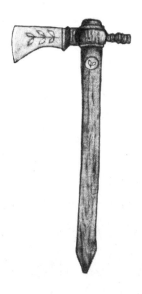

Tomahawk

1754, Western Pennsylvania

In execution, Henry VIII favored the axe. But on the battlefield, firearms had already started to take over by the Tudor era. Cannons, culverins, and other artillery were becoming more determinative than the pikes, swords, axes, and other bladed weapons still carried by most of the infantry of the day.

Then the gun got smaller, more convenient, more portable. Then it took over the world.

Firearms in the Seven Years' War were about as different from guns today as flint is different from steel. But from the beginning, their power to quickly and totally destroy a body was incredible compared with the swords and arrows and axes that preceded. In the 1750s, novel flintlock mechanisms meant that hunters and soldiers didn't have to carry around an already-burning match, nor did they have to wind a wheel before pulling the trigger; they just had to add a little powder and cock the gun. But getting the ammunition ready was still a multistep process that took precious time and could easily go badly.

The gun's power is great. But in 1754 that power was not accompanied by convenience, or easy access. The axe has always had that going for it.

The axe was a crucial tool in the English and French colonization of America. Much of Europe had been deforested by centuries of consistent occupation; early colonists came from cities where they didn't have to do a lot of the labor of basic survival. On this side of the ocean the forests were unimaginably thick. There was not just the usual work of chopping down trees for lumber; vast stretches of land had to be cleared so colonists could build their versions of habitation and agriculture.

The Indigenous people who were already living in North America had their own methods to build and grow, working with the bounty of the land instead of razing it. They had stone hatchets and other cutting tools, which they adapted to the availability of British steel, creating the tomahawk. This small axe was a symbol of war and diplomacy at once. And at a moment when tensions between European empires were high, after many years of Indigenous resistance to imperialism, it was the tool at hand one morning in a Pennsylvania glen, before the guns sounded, after the guns went silent.

SEVEN

YOU ARE NOT DEAD YET, MY FATHER

As a twenty-two-year-old, George Washington screwed up so badly that he set off an international war. But you wouldn't know it from his posthumous beatification. A bestselling biography mostly invented by a "broke itinerant parson with impeccable timing"[1] hammered the arrogance and inattention to detail out of his blundering youth with a little fable about honesty, cherry trees, and a hatchet.

"When George Washington was about six years old, he was made the wealthy master of a hatchet!," wrote Mason Weems in 1800, "of which, like most little boys, he was extremely fond. He went about chopping everything that came his way." Eventually, he came to a cherry tree that his father, Augustine, particularly loved, and he couldn't help but chop the hell out of it. And when Augustine came to confront him about it, the boy said, "I can't tell a lie, Pa, you know I can't tell a lie, I chopped it down with my hatchet."[2]

You are a smart person who reads contemporary nonfiction, so I'm assuming you know this is altogether untrue, but it's worth examining from the perspective of the axe.

Weems wrote this in a heavily embellished biography of the first president published soon after his death that quickly became a bestseller. The cherry tree story actually didn't appear in the book until a later printing, when it was already a hit. The lie about Washington's unflagging honesty became a part of American mythology when it was included in a children's reader and later appropriated by P. T. Barnum. His circus billed an enslaved woman named Joice Hath as Washington's nurse for a sideshow, claiming that she was over 160 years old; the cherry tree story was part of her repertoire.

What makes the story work is a little boy's overexcitement about a new possession, and the hatchet has an understandable appeal in the era before toy guns. For a child a hatchet would have the adult allure of responsibility, practicality, and danger—like a play kitchen with a gas stove. And even in a story about his childhood, Washington's masculinity is at the forefront. Here's a little boy who can chop down a tree all by himself.

In Grant Wood's painting *Pastor Weems' Fable*, George's hatchet is a fire axe, the red kind that wasn't invented until the twentieth century. It's iconographic, like much of the rest of the funny and layered painting: it's distinctly unreal, framed with Weems pulling back the curtain on a child with the face of Gilbert Stuart's portrait of Washington. Enslaved people in the background remind the viewer whose labor undergirded the myth of Washington's childhood hatchet job.

"Perhaps the tool least regarded"[3] but most necessary in colonial America was the axe. Prosperous London merchants like John Washington, George's grandfather, would not have been doing his own woodcutting in their home country; his strongest association with the axe could have been Tower Hill. But he would have had to use the axe for woodcutting as he cleared, surveyed, and planted his new land, as he and the people he enslaved tended to horse hooves and dressing flax. A great deal of steel had to be imported from Germany and Britain to meet the demands of those farms. And the axe held special value as a trading resource. Steel axes—often small-trade axes—were stronger, sharper, an immediate upgrade from flint. The steel axes were manufactured overseas, evolving "in response to indigenous demand."[4] British steel quickly melded with Indigenous toolmaking to create a new form of axe. The flared blade of the one-handed boarding hatchet used by the British navy evolved into a thinner instrument, less suited to woodcutting and more appropriate for war—and peace.

The tomahawk sometimes has a spike on the back of its head, but it is sometimes a pipe designed to share, to seal treaties and build relationships. The word "tomahawk" comes from the Powhatan word "tomahack" and the Abenaki word "temahigan," meaning "the striking instrument" made of thin flint and hafted. "Tomahawk is an indigenous word and as such has a special ability to evoke notions of bloodthirsty savagery," writes Dr. Scott Manning Stevens (Akwesasne Mohawk), Associate Professor of Native American and Indigenous Studies and English at Syracuse University. "The tomahawk remains the most Indian of weapons, even if not of our manufacture, an object introduced by Europeans and indigenized by my people ... So often we see it sink into the skulls of its victims: a tool transformed into a weapon, one that blurred the distinction between simple utility and excessive violence."[5]

The small hatchet was the "most feared weapon to the settler colonial imagination" but also a "powerful assertion of [Native] sovereignty when held in restraint."[6] Though the tomahawk was a symbol of butchery in much of the American imagination, it was also understood as a visual signifier of peace internationally: European portraits of Mohawk diplomats placed a tomahawk at the subject's feet to symbolize their peaceful intentions. This association was so much a part of Indigenous culture that our language still holds a place for it. "Burying the hatchet" is a reference to the Native practice of burying tomahawks or clubs. Still, "a Native offering of peace could be met with ferocity if any betrayal occurred."[7]

In his military career, John Washington was not known for his diplomacy. His nickname was Conotocarious, or Town Destroyer. Generations later, his grandson was honored to inherit the fearsome moniker, bestowed by an allied Iroquois leader by the name of Tanacharison (also spelled

Tanaghrisson), a man remembered for sharing the Washington willingness to strike first in the name of his people's life and liberty.

Tanacharison was a Seneca official who, in 1750, held great power in what is today Pittsburgh, at the fork of the Ohio River. He was a half king, from French *le demi roi*, a term of uncertain power in the Iroquois Council; it could mean great power over the conquered nations of the area, or it could be limited to relaying messages and gift giving, depending on the council's faith in the half king at hand. Half kings had power, but it was more representative and less executive: they could negotiate on behalf of the Iroquois but not make binding treaties. One of Tanacharison's duties was to make sure that the English and especially the French were respecting boundaries as the European nations fought among themselves.

There's a tendency to see the genocide of the Indigenous nations of America as fate, something that was always going to happen. It wasn't. In 1754, nothing was decided. The original people didn't have every advantage, but they knew and understood the land on which they lived. Native nations who had been in conflict for decades were finding ways to work together with and against the European colonialists trying to take as much of the country as possible.

The Great League of Peace and Power was the constitution of the Iroquois, also known as the Five Nations, dating back to before European contact and written in the form of wampum belts. Wampums were created painstakingly from shell beads and acted as the record and symbol of compacts between two parties; pictographs lay out the terms of the agreement and are used both for later reference and to signify the trustworthiness and weight of the person holding the belt. The Great Law of Peace, ratified by the wampums, was devised in part to address "mourning warfare," a form of perpetual blood feuding among the states involving raids and kidnapping.[8]

The Great Law of Peace did not eliminate internal conflict, but it allowed the Iroquois confederacy to grow in population and influence. In 1753, this made them the key to avoiding making long-term commitments to any other nation. They resisted any permanent concessions to European occupation of the land, preferring a more flexible arrangement that allowed them to trade with whomever offered them the best rates and quality of goods, often playing settler nations off each other. But as competition for the Ohio valley increased, that could not continue. The Mingos, Wyandot, and Delaware communities that had been subjugated by the Five Nations could look to the French for protection instead, which wasn't good for the half king.

As the French incursion into Ohio inched forward, a French commander showed up with a huge battalion. When the Iroquois confederacy sent a contingent of women to politely ask what they were doing there, the commander responded that he'd been holding a hatchet aloft so that everyone was clear on his intentions.

Tanacharison delivered to him a wampum belt along with a speech, in which he addressed the commander as "father" and yet clearly rebuked his attempt to encroach: "To come, Father, and build houses on our land, and to take it by force, is what we cannot submit to."[9]

In response, the commander threw the wampum on the ground. He called the Iroquois nations insects and said they had less land than what was under his fingernails. He told them they could expect kindness only if they would be ruled.

The Iroquois didn't want to throw weight behind one European nation or another. It benefited them to let the Europeans fight each other. But if they didn't want to fight the French alone, a deal with the English would have to be made. Tanacharison would have to ally himself with a very green George Washington, who seemed to immediately irritate his small crew.

Washington and Tanacharison's first journey together was to find and deliver a message to the French requesting that they give up their claim to the camps they were building in eastern Pennsylvania and summoning their captain to the English stronghold of Wills Creek on the Potomac. Their crew followed Tanacharison's timetable, taking detours to pick up wampum belts without which their diplomatic "words carried no weight."[10] Once they reached Fort Le Boeuf (modern-day Erie, Pennsylvania), the seasoned French captain deftly outmaneuvered the solemn twenty-one-year-old, avoiding any commitments and preserving the relationship with Tanacharison. Washington was reduced to an "errand boy," sent back with a letter in tow.[11]

His journey back in the dead of winter, without his Indian guides, was nearly fatal. They tried to travel by the frozen river, constructing rafts with just "one poor hatchet" between them. Washington was sent out with two hundred men to the forks of the Ohio to build a fort. It was a miserable winter of pointless work. There were not enough supplies or food, and Washington felt he wasn't getting paid enough (though he still complained about his Indian allies wanting to be compensated for their assistance). They certainly weren't willing to help on Tanacharison's word. Even though he tried to sell it to his fellow Iroquois leaders as a positive development, they didn't buy it and wouldn't help the deployment with food, not even if well compensated.

When the French delegation arrived at the makeshift fort, they were amused by Washington's silly idea of a fort. They had twice as many men as the English and they were going to knock down their amateurish little structure and build a real fort. Their only concession was to treat the men still working there to a nice dinner because they'd been so accommodating about the whole thing.

Tanacharison, however, was extremely pissed off. A French victory meant his loss of influence. If the French gave shelter and protection to some of the lower-ranking tribes of the area, it reflected poorly on Tanacharison's ability to keep them under the Iroquois confederacy's suzerainty. Not only did the French scoff at his protests, but his Shawnee and Delaware peers ignored him too. Tanacharison was not one to take these losses in stride. As the Virginians quietly filed out of the fort, already in the process of being torn down, he stayed behind to "[storm] greatly against the French."[12]

England and France were technically at peace at the

moment, so the French could swagger and stalk and threaten and intimidate, but they couldn't really use force. As Washington and his soldiers made their slow and noisy way southwest, the job of tracking them fell to Ensign Joseph Coulon de Jumonville. He was the "scion of a distinguished military family,"[13] but his force was small; they couldn't do much more than spy and maybe politely ask what the British were up to.

Tanacharison, far better at tracking, was intent upon following the French. He sent Washington an alarmist message about a fictional potential attack with the location of the French unit—right after Washington had sent most of his men in exactly the other direction. Washington set off in the rain at 10 p.m. with forty-seven thoroughly exhausted men in tow, reaching the half king's camp around daybreak. Tanacharison and Washington had a brief council, during which the older man manipulated the younger into a surprise attack against the French.

Jumonville and his men were camped about fifty miles southeast of the fork of the Ohio, in a lush glen at the bottom of a small fern-laden valley. On the morning of May 28, 1754, after a few days of heavy rain, it would have been nearly steaming with precipitation.

The French battalion was just waking up. There were no sentries posted. Washington told the Native contingent to cut off escape routes and ordered his men to surround them quietly, but some kind of noise—perhaps a musket misfire—got the French wide awake. They scrambled for their guns, which were too damp to properly work, and Washington gave the word to fire.

All was gunpowder for ten to fifteen minutes. The French tried several times to escape but were driven back from the ravine nearby into the field of conflict. Jumonville was finally able to call a ceasefire to try to communicate with Washington. He had a peaceful summons to deliver from his commanding officer, a message that indicated the French's intention to "maintain the concord which now reigns between our two friendly princes."[14]

Washington couldn't understand the message, as it was in French, but Tanacharison did, and he was not inclined to take Jumonville in good faith. Jumonville himself had done nothing wrong to him, but this wasn't about Jumonville personally. This was a dispute with the French authority in general.

"Tu n'es pas mort encore, mon père," Tanacharison said in French, the language of diplomacy. "Thou art not dead yet, my father."[15]

Then Tanacharison pierced Jumonville's skull with a tomahawk, surely above the hat-brim line.

When that was done, he reached into Jumonville's skull cavity and grasped the dead man's brain: "Because the gray matter would have been the consistency of thick, wet plaster, the half king could, in fact, have squeezed it between his fingers, seeming [to] wash his hands in the tissue."[16]

That is not the story Washington told to his superior (after complaining extensively about his pay) or his diary, which was more a recitation of the most basic facts: Jumonville and nine other French dead, passive voice; twenty-one captured, including the second-in-command; the Native warriors took the dead bodies and scalped them. Washington was already a sober, serious fellow, savvy enough to not take credit for the win or the mess. He was certainly no elder statesman. He had never seen any kind of real violence; he thought the whistle of bullets was "charming." So he set up a circular little "death trap" fort in a terrible tactical location, even though Tanacharison told him not to.[17]

But he did know that he had won, and also that he'd lost control of the situation, failing to realize that Tanacharison might have a different agenda.

The new leader of the French force, Captain Louis Coulon de Villiers, was Jumonville's half brother, and unlike Washington he was not learning on the job but a veteran of twenty years. Plus, he came with considerable reinforcements from French strongholds in the north. And he had allies among the Mingos, Delaware, and Shawnee, who were trying to get out from under Iriquois council authority and "take up the hatchet on behalf of the French."[18]

About five weeks after Jumonville's murder, Villiers and company arrived at the meadow near Fort Necessity. They occupied the forested area slightly above the battlefield, which offered a ton of tactical advantages: they and their firearms were sheltered from the rain, and they had the upper view of the land.

Tanacharison repeatedly urged Washington to surrender, and left when the young man didn't listen. The English were in the rain and muck, their guns useless. The enemy was at a far enough reach that the English couldn't use a handheld weapon, like a dagger or a hatchet. And they were exhausted and in many cases sick after the stress and toil of the last few months. The minute the sun set, the English decided to give up and go get drunk. They figured they would be slaughtered soon anyway, so they understandably broke into the rum.

It was not an easy hour for young George Washington. His first skirmish had gotten out of hand immediately, and now this one was also going very badly on an even bigger scale. This was a confident young man, tall, handsome, and raised rich. He'd already started a successful surveying business. Now he was finding that managing supplies and men and enmity was a great deal harder than he thought.

And then an out appeared.

Villiers offered him a compromise. If they capitulated, gave back prisoners, and signed an agreement to leave the Ohio valley, they could leave with their possessions and honor intact.

Oh, and one other little detail: they had to take responsibility for Jumonville's assassination.

Washington was working with a translator of questionable quality (or perhaps just a touch of illiteracy), and the French were drawing up the documents he signed. He knew about the capitulation, the leaving of the Ohio, all of that. But Washington and his advisors absolutely did not know they were basically pleading guilty to the French king. So he signed the agreement, thinking that it was a failure but not the beginning of a war.

How wrong he was. It took another year and a half for war to be declared, but the conflict soon spread worldwide, lasting seven years and encompassing most of Europe, South America, India, and the Philippines.

The war itself wasn't his fault, and Washington was given an impossible task. But he chose the wrong place to make camp and do battle. He failed to realize what Tanacharison was doing, leading him to the French that day in May. Yet George Washington failed up: his "disastrous performance on the frontier somehow turned out to be a social climber's dream."[19] The governor who made the foolish choice to send the inexperienced young man into battle had Washington's journals published, and they were a hit. War stories are great propaganda, and violence is eternally saleable.

Tanacharison's motivation for his vendetta against Jumonville and the French in general is as awful and brutal as his violence. In his telling, a French delegation killed and ate his father when he was a child. Boiled him, to be specific.

This story is likely not about Tanacharison's literal father but another man, Memeskia. He was a chief of the Piankashaw and the primary voice of resistance against the French. In 1752, a Métis man named Charles Langlade led a French-affiliated contingent of Ottawa and Chippewa in a siege of Memeskia's town. They quickly won and decided to send the message home with some ritual sacrifice.

First, they killed a trader and ate his heart. Then, with his people looking on, they "killed, boiled, and ate Memeskia."[20]

Like Memeskia, Tanacharison was a leader, a proud man who was devoted to resisting the French. Most of all, Tanacharison wanted to recover his status and his power over his fellow people. What mattered to him was not so much the conflict among European nations, which he had tipped into war, but the esteem of his own nations. But, as he knew, he put his trust in the wrong person with Washington.

Still, he worked to keep Native support for the English. When the Delaware and Shawnee defected anyway, it was a sign that the reach of the Iroquois council was weakening.

The half king was broken, heartsick, and physically unwell. He collected his family and friends and decamped to south central Pennsylvania. A healer diagnosed him with having been bewitched by the French in retaliation for Jumonville (though it was probably pneumonia). He died in October of that year.

At the National Parks site for Fort Necessity, there were several axes among the leggings and muskets on display. The most striking was not made of metal but shells, a beaded silhouette of a hatchet on a belt. It was a reproduction of a wampum that circulated to spread word of an impending conflict, composed of shell beads dating back to the time of Tanacharison. The background was red, representing diplomacy and friendship, surrounded by black beads symbolizing death. In the middle is the hatchet in white. The axe is the war.

War, like execution, is a collective act, an act of the state. In this way, it is distinct from murder of an individual due to interpersonal conflict. Tanacharison did not negotiate or observe the usual terms of engagement when he killed Jumonville; he didn't hold the power to make that kind of statement anyway. Jumonville's killing was not performed on behalf of the Iroquois council or the English. It was personal. And yet. The goals Tanacharison wished to accomplish by striking an unsuspecting man in the head and reaching in to grab his brains were not the petty motivations of Freydis or Henry VIII. His act was not careless, nor an act of tyranny: it was an act against tyranny, more like the force animating Stesagoras's assassin more than two thousand years earlier.

That day in the glen occurred after a long period of French and English forces muscling their way into occupied land. Europeans were perfectly willing to use equally twisted and excessive interpersonal violence to attain their goals. Indigenous nations were certainly capable of great violence too but they were also practicing a sophisticated system of government with well-established rules for resolving conflict. It's just that these strategies were of limited usefulness when the European occupiers considered them optional. The metal that transformed the tomahawk Tanacharison used that day was an adaptive response to the changing face of the old nation. In the face of guns and germ warfare and European arrogance, resistance could not always be nonviolent.

Tanacharison's violence was his own choice, uncoerced, ordered by no one. He did not have to kill this man in this way on this day. Yet I see it less as murder and more as an act of war in response to an ongoing genocide. The Seven Years' War had yet to be declared, but the campaign of violence against Indigenous people had already lasted centuries.

Cooper's Side Axe

1861, off the coast of South Carolina

The axe was out of favor as an implement of state violence by the nineteenth century. Guillotines were the stylish way to decapitate someone after the recent French Revolution, but American capital punishment never tended toward beheading. Hanging was the predominant method of execution in the states, and the firing squad was another frequent choice. Firearms were the preferred form of officially sanctioned violence, not just in the military and death penalty but in the earliest formal iterations of police forces: slave patrols.

Enslaved people had so many various forms of labor to

perform that it would be inaccurate to say that the axe was particularly distinguished among their tools. But even if it didn't stand out, it was still omnipresent. There were many specialized axes on a plantation, all sharp enough to be used in violence (like the tobacco hatchet, a one-handed tool with a short, squarish head specifically used to cut the tobacco plant at the base before it was hung to dry). Nat Turner's rebellion was armed with a multiplicity of axes; Turner "spill[ed] first blood" with a hatchet and his compatriot Will acted as "the executioner [with] his fatal axe"[1] as they sought "dear liberty"[2] in Southampton.

But as the axe continued to grow still more common, it was even less likely to be noticed and interpreted as a threat, especially in the hands of a workingman. By 1861, the development of the Bessemer method of quick and easy steelmaking had led to scores of manufacturers of axes, many with their own signature patterns and knockoffs of other patterns. On land and on sea, the axe was an "everyman's status symbol," as described by Brett McLeod, forestry professor at Paul Smith's College and author of *American Axe*, in a 2023 interview.

As we learned with the Vikings, axes are useful on a ship. In 1863, they were primarily in use to handle the boxes and barrels of the cargo. Like ships, the wood of barrels needs to be very precisely hewed, and may need quick repair after the damage that comes with a tossing and turning ship. A hatchet is small, handy, and distinctly wieldy for cutting ropes or meat or throats.

EIGHT

I SUPPOSE YOU KNOW WHAT I AM DOING

William Tillman was just fourteen years old when he took to the sea. He was born to a free black family in 1834 in Delaware, the northernmost slaveholding state, where the Tillmans were limited in their employment options to "farmhands or domestic servants."[3] There were opportunities for a young black man on water that could not be found onshore; this was pre-1850, before railroads totally dominated America, and when a lot of travel, transport, and shipping happened by water.

So the Tillman family's move to Rhode Island, the "ocean state," probably suited him just fine. Rhode Island was one of the few places where free black people could legally move, where they had a right to be represented in court and build a business. It was also a very white state, where free black people were treated with fear and distrust, both in the eyes of the law and in daily life. In New England, "free blacks most often worked as cooks and stewards, positions that whites were loath to occupy [because] they were not deemed manly."[4] But on a ship, being a cook brought authority and power over the galley that only the captain could best.

From 1856 on, William worked as a cook and steward on a 119-foot ship named the *SJ Waring*. On July 4, 1861, they were setting off from New York to Uruguay and Argentina on a commercial voyage, transporting cargo worth $100,000 in 1861 money. There were ten men on board, including an Irish passenger named MacKinnon and a German sailor named Billy Stedding. They left behind a country in the grips of a new war. William had been to southern ports on the eastern seaboard, and had seen enslaved people laboring and being sold. He was not going to have any part of that. He was headed to another continent.

Until he got to New Jersey and the war came to find him.

The weather was good and the voyage was going well on the morning of July 7, 1861. The *Waring* was just about two hundred miles "to the south and east of Sandy Hook, the barrier peninsula that guards the entrance to New York Harbor." It was 10 a.m. when they spotted a ship on the horizon, a black brig flying French colors. The first mate didn't like the look of it, but the captain refused to alter their route.

Once it was too late to change course, the other ship pulled off the mask. As soon as the *Waring* got close enough to see the ship and their guns clearly, they fired a warning shot. It was the USRC *Jefferson Davis*, a schooner filled with Confederate pirates.

The Confederates ordered the *Waring* crew to lower their flag. The *Jeff Davis*, as it was called, had already taken two ships on its expedition to make trouble for the Yankees. As it happened, the captain of the *Jeff Davis* knew the captain of the *Waring*, and the capture of the ship went pretty smoothly; they even had brandy together while the pirates stripped the place of firearms and many perishable goods.

It only took about an hour. The pirates brought over most of the crew to their ship. Much of the cargo they left intact, to be unloaded when the prize crew they installed had sailed the vessel back to South Carolina. MacKinnon, Stedding, and Tillman remained onboard the *Waring*, tiptoeing around the prize crew. Tillman continued to expertly perform the duties of the trip, cooking and serving food, cleaning, and doing what was asked of him. He never spoke unless spoken to and he addressed his captors with the same respectful formalities he would have his captain.

The moment he decided to kill them was born quietly too. After the flag was lowered, the privateers ripped it apart and refashioned it into that enduring symbol of hatred, the Confederate flag. Tillman watched as they sewed the scraps together into "what passed for a rebel insignia" and silently stewed. The flag was what William later described as his first "incentive for revenge."[5] First but not greatest.

"The pirates had chuckled over [their] good luck," wrote a report on Tillman in Frederick Douglass's *Douglass Monthly*.[6] They were planning to sell Tillman—whom they knew to be a free man—into slavery.

The prize crew were not subtle about their plans, openly

speculating about whether he would bring in over a thousand dollars at auction. "By God, he will never see the North again," one member of the prize crew said.[7]

Before the captain left with the *Davis*, he called Tillman to him and told him to report to his house in Savannah to be "take[n] care of." In Tillman's telling, he tipped his hat and said, "yes sir."[8]

Tillman had to keep his countenance cool, so his captors would relax around him. But he said to Billy Stedding, the German *Waring* sailor, "I am not going to Charleston a live man. They can take me there dead."[9]

Tillman hoped for rescue by the United States Navy for a while, but eventually those hopes faded. He began to plot a mutiny with Stedding, who as a German citizen could well be in danger of becoming a prisoner of war in Charleston, or worse, being conscripted to serve in the Confederate army. The other member of the *Waring* crew, MacKinnon, had befriended their captors. They also approached a Scottish member of the prize crew, but he refused to join them, though he said he wouldn't try to stop them. Stedding and Tillman had to somehow kill or detain the other four members on their own.

Luckily, Tillman had stowed a hatchet in a corner of his room.

The hatchet was "an innocent instrument," wrote Brian McGinty in his excellent account of Tillman's escape, *The Rest I Will Kill*. It was used "for routine tasks on the vessel, like trimming rope, cleaning fish, or cutting wood for the cooking fires." Enslaved people undertook all of these chores and more, and so it often made sense for them to be carrying a hatchet, even when forced into servitude. Tools would often be housed in enslaved persons' quarters, as it was usually their responsibility to keep them in good repair.

In other words, having an axe in their berth was something you'd totally expect from a person whose job was to make the food and keep the ship clean and comfortable. It would have been fine to be caught holding an axe. But Tillman, like an assassin in Herodotus, kept it concealed anyway.

On the night of July 16, the *Waring* was beginning to approach Charleston through a series of inlets and backwaters. Tillman went to his room as normal. Stedding was at the wheel; he was supposed to give a small cough as a signal for Tillman to attack, but Tillman fell asleep or didn't hear it. Stedding went quietly to Tillman's berth; all of the doors had been taken off to improve circulation in the muggy air.

Stedding put a hand on Tillman's shoulder. "Now is our time."[8]

Tillman went stealthily to the captain's room and hit the man twice in the skull with full force. On the other side of the cabin, MacKinnon heard a faint scream and saw Tillman cross to the second mate's lodging, "striking a kind of sideways blow."[11] The man tried to defend himself but soon fell to the floor. MacKinnon and Tillman made eye contact.

"Sir, you needn't be scared," Tillman said. "I suppose you know what I am doing."[12] MacKinnon agreed, and survived.

Stedding met Tillman on the deck, where Stedding held the first mate at gunpoint. They dared not shoot for fear of waking the other Confederate soldiers, who would still outnumber them. Tillman used the hatchet on him instead. They pitched the wounded man overboard, and Tillman went back to finish off the injured men in their room. They went overboard too. The rest of the soldiers were finally awake, but too stunned to do much of anything. Stedding and Tillman got them in chains pretty easily.

"We have done all the butchering I guess we will do this voyage," Tillman said. "There is two of them and there is two of us. I guess we can manage them pretty well without taking their lives at any rate."[13]

After ten minutes of horrible violence, Tillman was fully in charge now. He tried to say that Stedding was the leader and insisted on continuing his duties as cook and steward, that he wouldn't be the captain. But the more they discussed it the clearer it became: the man who would not be a slave was the one whose instructions they would follow.

Only problem is that no one on the ship really knew how to sail.

There were only six people on board, which was barely enough to keep up with the basic functions of the boat. Tillman and Stedding had to remove the shackles from the Confederate men and hope they would work to keep the ship on course; perhaps the late captain and mates were not well beloved by the leftover pirates. Of course, figuring out that course was a huge task, one that could go very badly. Just figuring out the basic direction to go was difficult, let alone managing winds and tides. MacKinnon was literate and was able to read guides and almanacs to help avoid dangerous areas like the Outer Banks of North Carolina.

Primarily, they relied on Tillman's instinct, which proved to be wise. Nearly twenty years of experience on the sea taught him how to stay close enough to the coast not to get lost while dodging the privateers and pirates who were looking to pick off inexpertly piloted ships.

Five days after he threw over his captors, William Tillman guided the *SJ Waring* ashore on Fire Island. A pilot boat brought them to New York Harbor. It was known that the *Waring* had been taken, but not that the captors had been overthrown. The return of the ship and its cargo was a cause for celebration.

Tillman's brave self-rescue became national news. The coverage was, of course, racist, with patronizing descriptions from publications like the *American Phrenological Journal*, and a lot of credit-to-his-race type of rhetoric in mainstream reporting. But it also portrayed Tillman as he was: an exceptionally brave man and an expert sailor. With the war building to a full boil, Tillman was a demonstration that black people did want to be free, that they had the same capabilities as white people, that they loved this country too. In Frederick Douglass's periodical *Douglass' Monthly*, Tillman is a "black hero." His "fearful work [with a] common hatchet" is a demonstration of the best the United States has to offer, and of why the Southern Democrats who insisted on concessions to slavery could not be mollified. "Love of liberty alone inspired him and supported him, as it had inspired [other] negro heroes before him," the reporter wrote, citing his antecedent in seeking freedom with a hatchet, Nat Turner.[14]

By saving the highly valuable ship and its highly valuable cargo, Tillman and Stedding were due some money. The ancient rite of salvage says that "those who saved vessels and their cargoes from losses at sea were entitled to be rewarded for doing so."[15] So Tillman and others sued to be paid for their efforts. Because his own welfare was tied up in the welfare of the cargo, insurers claimed that he lost that right to compensation. The argument was that since he was fighting to stay free, his rescue of the ship and its goods was just a nice windfall for his employers.

The argument didn't float. Tillman was awarded seven thousand dollars—hundreds of thousands in today's money, and the most of any of the men on the voyage.

After his brief burst of fame, the rest of his life was lived quietly and privately. He registered for the draft but doesn't seem to have served for the Union. He married a woman named Julia and had at least one child while living in Rhode Island. He might have moved to California before 1880, when a William Tillman with a sore eye appears in the 1880 San Francisco census. He was still a cook.

William Tillman and Billy Stedding were ready to die that day in July 1861. The prize crew of Confederate soldiers had already raised the hatchet of war. Tillman and Stedding acted in self-defense. The hatchet Tillman hid in his room can be seen less as a weapon and more as a tool. His purpose with it was not primarily to kill, but to survive.

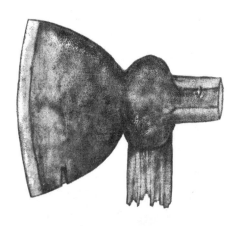

Roofing Hatchet

1892, Fall River, Massachusetts

In 1892, people had a lot of axes around for a lot of tasks. But the brand-newness of readily available and inexpensive steel tools had worn ragged in the presence of trains and skyscrapers and factories and other new technology made possible by steel. Having a lot of hammers and hoes and hatchets around was convenient, yes, but it definitely decreased prestige. Axes were not novel except as actual novelties, an echo of the tiny axe tokens in Ahhotep's tomb. At the Chicago World's Fair, vendors sold tiny glass hatchets emblazoned with George Washington's face; they were a hit, reproduced for decades after in tourist traps and the like.

Carrie Nation had this instinct too. In the 1890s, the temperance movement made a habit of "saloon attacks" executed by "women singing and carrying axes."[1] They didn't specifically use axes—smashing up bars with rocks and pokers and hatchets—but Nation, a leader in the Woman's Christian Temperance Union, rose to national prominence as a saloon attacker once she made the hatchet into her personal emblem. After she became famous, she too made merch out of tiny axes, selling tiny hatchet pins with the word "home defender"—underscoring the axe's legacy as a feminine symbol of hearth and home.

But even as axes became more gimmick than essential tool, they retained their menace in the growing newspaper industry. Every murder weapon has an innate interest. Helen Jewett, murdered in 1836 "at the business end of a small hatchet,"[2] was found buried in the backyard of the brothel where she lived. The story of her murder and the trial of her ex-lover Richard Robinson fueled competition in the recently minted trade of cheaply printed penny papers. One of the editors of these papers claimed that the axe specifically pointed toward "the vengeance of female wickedness"—and that a more masculine method would be dagger or poison.[3] Robinson was eventually acquitted.

When Helen was murdered, the newspapers could use the axe to sell the brutality of her crime. Yet for much of the nineteenth century, when the words "axe" and "murder" appeared together in a newspaper, it was an accident of adjacency. "On the night of the 27th of January last, Mrs. Brita

Nelson, with an axe, murdered her husband, Peter Nelson, at Rapids City, in this county, a coal miner, while he was asleep in his bed," as the *Chicago Defender* happened to put the words together to describe an 1876 butchery.[4]

Washington Frothingham was the first to put the phrase together in 1877. A writer, philanthropist, and reverend who once ministered especially to lumberjacks, he wrote a syndicated column for a number of newspapers in New England under various pen names. To the readers of the Rochester, New York, *Democrat and Chronicle*, he was Macauley; in Middletown, Connecticut, he was the Rosicrucian; and to the people of Rutland, Vermont, he was the Hermit of New York. In one of his columns as the Hermit, he detailed several incidents of spousal murder by axe from the 1860s.

"These ax murders were for a time frequent, and what seems more horrible, they were the deeds of husbands," he wrote, perhaps coining the compound phrase. The common beats of domestic violence assume that the murderer is a man and the victim a woman. A woman wielding the power of the axe against the man of the house was an irresistible thrill to Victorian-era true-crime fans.

NINE

FIVE AXES IN THE CELLAR, ONE AXE ON THE ROOF

Throughout the coverage of Lizzie Borden's trial, the newspapers called the purported murder weapon "the hoodoo hatchet."[5] It was a small roofing hatchet with a broken handle, an ordinary object with no overt religious symbolism. "Hoodoo" had a few definitions at the time: sometimes it referred to a raucous party or a vigilante mob, and sometimes it was a racist epithet. Most frequently, it was a derogatory way to refer to the sort of spiritualism that was very in vogue in the 1890s, an attention-grabbing synonym for everything from witchcraft to augury to superstition. The hatchet's small size and lack of handle may have suggested a charm, a religious symbol of life and death and power, just as it did for Viking kings and Chinese emperors and Roman dictators and pre-Neanderthal mourners.

"Hoodoo hatchet" was ultimately a way to call the famed axe bad luck. The New York *Evening World* positioned the Borden hatchet head against George Washington's hatchet from the cherry tree story, "the emblem of truth." Her broken-handled hatchet was "the antithesis of the Washington hatchet: it cannot tell the truth."[6] So it was known for its "demonish pranks" against the prosecutors when they brought it to trial.[7]

There were three hatchets in the basement of the Borden home. Two were suspected murder weapons: one with a loose but intact handle, another with a handle apparently tightly hung but cracked and broken. Both were found so wanting that, by the end of her trial, her defense was bringing these hatchets up more than the prosecution was.

I don't know whether or not Lizzie did it. I understand why people think she did: there was so little time, so little opportunity for anyone else to sneak through the house and kill two people in the space of less than two hours. And she had motive enough. But if there was too little time for an assassin or serial killer or some other fiend to slip in and out, there was equally little time for Lizzie to silently kill two people, change her clothes, clean herself, and hide the weapon so well it was never found.

The overworked Borden housemaid Bridget Sullivan was up first on August 4, 1892; at 6:15, she headed down from her attic room to the cellar to get wood and coal, fuel to make the hot day hotter. Andrew's brother John was staying with them, and he was up not long after, hanging out in the sitting room where Andrew would soon die.

Lizzie's stepmother, Abby, was the next to rise. She married Andrew when Lizzie and her sister, Emma, were young children, but her relationship with her stepdaughters was always tense. Andrew came down and did a few chores in the yard and barn before sitting down with his wife and brother for breakfast. After they finished, Bridget sat down to eat while the other three chatted for a bit—though Abby soon got up to do some dusting.

Lizzie came downstairs around a quarter of nine for breakfast. Maybe it was a cookie or a banana—she's vague about these details at the inquest later, as she was on many topics. On August 4, she was in the middle of her period and didn't feel great, but she set about the task of ironing the handkerchiefs.

Her uncle John had left the house to run a few errands and visit family, expecting to come back since Andrew had invited him for dinner on the way out. Abby was dusting, and Andrew was getting his collar and tie ready for the day ahead. Bridget had been told to wash the windows, but she was outside vomiting. The lamb stew the household had for dinner the night before had gone bad, but they kept eating it, such was their attention to the budget.

At nine, Abby went upstairs to put shams on the guest room pillows, and Andrew left for the post office. Sometime between nine and ten, Abby was murdered in the second-floor guest room.

Bridget, having recovered from her upset stomach and set to her task, saw Lizzie at the back door at 9:30 and told her that "she need not lock [the] door."[8] While Lizzie waited for the iron to heat up, she said, she leafed through an old issue of *Harper's* magazine.

Andrew was seen leaving the shop sometime after 10:25. At 10:40, Bridget let Andrew into the house. For some reason the door was triple-locked, and Bridget fussed with it. They heard Lizzie laughing at Bridget's "pshaw" on the stairs. She came down and had a quiet word with Andrew.

Andrew went up to his room on the second floor, taking a route that did not pass his wife's cooling body. He came back down and took a nap on the couch at a quarter to eleven. In the kitchen, Lizzie asked Bridget a few questions—about a note Abby was to send someone, concerning a sale.

Bridget went up to her room to take a nap too. She heard the City Hall clock chime at eleven.

Lizzie says she wandered out to the yard, enjoying a ripe pear and looking around the barn for some fishing equipment for an upcoming excursion with friends. She rarely went in there. More bad timing, if it's actually where she went. Plenty of people saw Bridget going about her business when she was washing the windows; she chatted with a neighbor, even. People saw a man taking pears from the yard from over the fence right around when Andrew was settling down to his final nap, and a buggy pass by around then too.

But no one saw Lizzie in the yard enjoying pears. Her whole inquest is like this—foggy on the details of that morning, ambivalent about her feelings toward her family members, unable to explain her unusual activities. She says she heard a scraping noise during this time period, and turned to see the side door wide open.

At ten minutes past eleven, Lizzie called for Bridget—by shouting "Maggie!" as the Bordens always called her by the name of their previous maid. Andrew was dead.

Bridget and Lizzie hurriedly called neighbors and friends for assistance, and a police presence was there within ten minutes. Within a half hour of Andrew's death, a dozen people had seen Lizzie, wearing the same dress Bridget saw her wearing that morning, looking as neat and unbloodied as any good upper-class menstruating Victorian lady. Like she could not have hacked her father to death less than an hour ago.

In the photos taken shortly after his death, Andrew looks natural, unposed, his hands lying on his abdomen. A normal guy taking a nap, except for half his face being gone. Whoever killed Andrew committed the act while standing behind him, probably trying not to wake him while he slept on a cushion and a rolled-up jacket. The blows he received were a little less relentless—there were only ten of them. But there were enough of them to take off half of his face, crushing the left side of his temple. There's a sense of trying to get it done and over with as quickly as possible. The killer knew as well as we do that they had a very small window of time within which to accomplish their goals.

On the scale of this book, the Borden murders are very recent, modern enough that photographs of the scene are available. But they were long enough ago that the seemingly obvious idea of preserving the crime scene as it was found was not a priority. These were the days when passersby would simply be allowed into crime scenes. A lot of forensic advances, even recent ones like blood spatter, are downright stupid, but the idea that a crime scene should be preserved in as great detail as possible is a good one, and we weren't there yet. And so the photos of Abby's corpse were weird, the bed and other furniture moved to allow for a better view of her destroyed head—to little effect. There's not much you can tell from it.

Of the eighteen total wounds to her head, thirteen went through her skull. There were "fourteen parallel blows" to the right side of her face, caving a hole you could throw a softball through in the side of her head.[9] It was a passionate attack, possibly from behind due to slashes to her neck and crown, and the angle suggested that Abby's killer had straddled her body—an unthinkably unladylike position by Victorian standards.

When their bodies were first examined, it was clear to the medical examiner that Abby had been dead longer; for one thing, she was cool to the touch, whereas Alexander was warm. Her blood was "ropy," dark and matted on her head, whereas Andrew's was still "oozing."[10] When he examined their stomachs, it was clearer still that she had died first, as Andrew's breakfast was fully digested and hers was not. There was no sign of poisoning.

The medical examiner decided the weapon was a hatchet or cleaver right away, because of the force and the sharp edge of the cuts. By the time Andrew's and Abby's bodies were properly autopsied after going through the process of burial and exhumation, the damage to their heads was too extensive to tell "how sharply the edges had been cut."[11] They were about as decomposed as you'd expect of dead bodies left to sit unembalmed for a week in August. Near Abby's right ear was a "small but unmistakable deposit of the gilt metal with which hatchets are ornamented when they leave the factory."[12]

But the hatchets and axes in the basement were far from new, covered in grime and dust and ash. Lizzie went out of her way to make sure Bridget showed them to the cops.

The Bordens were living below their means. Andrew was a rich man, and this was a solidly middle-class neighborhood, lots of doctors and lawyers. But the house didn't have the updates becoming common at their neighbors' houses, like gas and running water. Andrew would rather buy a cheap hatchet and try to do his repairs himself rather than paying someone to appropriately renovate the house. There were a couple of working sinks in the house, but the only flush toilet was in the basement—where they kept the axes and the wood that the household required. The house ran on wood fires for heating, cooking, and cleaning—all constant activities in the Borden household, since they didn't have much staff beyond Bridget. Bridget had a lot of reasons to go down to the cellar.

She didn't do the labor of cutting down trees for firewood, of course. The work of turning them into logs, scoring the wood using the hatchet head as a guide and sharp edge, using a wood club to pound it in further, inspecting it for greenness and "smear[ing] it with fat [to] keep it from cracking"—that was all someone else's job.[13] A farmer came to cut the wood down to manageable size, usually once per winter. But they had skipped the previous winter. Once the wood was at the house, managing their wood supply was probably a big part of Bridget's morning chores.

But she wasn't the only one down in the basement that day. Lizzie was down there, attending to her period; she and her doctor both said that she was menstruating that Thursday. In the trial, Lizzie's period was described in the "possibly unique" Fall River parlance as "flea bites."[14] Menstrual blood was even more inappropriate than murder blood to Victorian minds. But it was vital to a violent story covered in gory detail. It goes to two specific elements of the investigation.

First was a speck of blood about the size of two fleas, or a "grain of rice," on her underskirt, which is honestly pretty tidy.[15] But Lizzie was a tidy person. The cleanliness of her dress is as crucial to the mystery as the weapon. The clothes she wore that morning were spotless except for that tiny bit of blood. She was seen by Bridget after Abby's death and again by many more people just twenty minutes after Andrew was last seen alive.

There are explanations for how she could have killed two people at close range without soiling her dress; it is a fact that Lizzie burned a dress (a common method of disposing of old clothes) later that week. Still, two flawless quick changes of a Victorian outfit in a short period of time is not a trivial matter; it's certainly not unthinkable that she committed the murders while maintaining spotless garments, but it's a strain for me to imagine.

Second, there was a bucket full of bloody rags in the basement.

These could have been a set of rags used to hurriedly clean an axe. Or it could have been a lot of pads. In the 1890s, women wore napkins "slightly smaller than a diaper" made of "bird's eye" cotton, rinsed and then soaked in copper.[16]

The number of axes in this basement shifts and expands like an old wooden door in the extremities of winter and summer, and I have trouble getting it open; I feel like I need an axe just to force my way in.

Here's Bridget, here's Lizzie, here's the police and doctors and neighbors coming through, failing to resist the urge to move a bed or a hatchet around. Even in the transcripts of the inquests and the trial, there seems to be little agreement on the number of axes, and a great deal of back-and-forth over whether there were two axes in the wash or on a shelf in the vegetable cellar or by the chopping block. This is complicated by the fact that some of the officers found it hard to distinguish between axe and hatchet. In some accounts, there are four woodcutting axes and three more hatchets, not all from the loot box in cellar (the box of iron tools that Bridget made sure to show the police).

I believe there were a total of five axes in the basement: two woodcutting axes, two usable hatchets, and a handleless hatchet.

No one thought that Lizzie used a woodcutting axe. An axe requiring two hands takes strength, coordination, and practice to use accurately; most people are more likely to hurt themselves trying to swing it than to hurt anyone else. It's possible even for a small, rather awkward woman like Lizzie to wield such an axe, but not with the efficiency, silence, and speed necessary for this specific crime.

A hatchet is an axe. But as a hatchet is not for felling trees, the woodcutting axe is not for every murderer. Hatchets are light and sharp, making it easy for anyone with basic hand-eye coordination to put their whole body weight behind. The medical examiner said that whatever the weapon was, it was something with a sharp edge and the leverage of a handle. A small handle and a lighter head make it easy for a petite woman like Lizzie to use—but no one thought she could effectively swing a woodcutting axe. So the focus from the start was not on the two two-handed axes but on the hatchets.

When police first examined the hatchets on August 4, they clanged against each other loudly—which is a detail that suggests there were two hatchets with handles in the box. The sound of clashing metal caught everyone's attention—especially Lizzie, who looked over with interest. I don't think this is particularly significant; anyone's nerves would be on edge in this situation, and I don't see how someone reacting to a loud sound is particularly exculpatory or not, but I'm duty bound to call attention to all the relevant axe facts here. She would later explain that she knew nothing of the axes, but she thought her father used them to decapitate pigeons sometimes—the kind of small task he wouldn't hire out.

Officer Fleet (great name) found these hatchets in this box. One of them was a shingle hatchet—a specialty tool with a longer and more rectangular head with notches for measuring shingles. These had a straighter cutting edge than most axes, and a blunt hammer. The other was a claw-hammer hatchet, about four and a half inches. Very sharp. Very clean. With one red spot on the side. And is that a hair?

A claw-hammer hatchet is a perfect kind of axe for exactly this kind of crime. Unlike the club Lizzie's friend Alice would find in Abby and Andrew's bedroom the next day, a hatchet has a powerful, sharp edge. A shingle hatchet would work, but the length of the head and the perpendicular angle of its bits (the bottom and top corners of its blades) would have been awkward to handle. But the claw head is especially handy for the purposes of murder. An axe head is usually quite blunt, but a claw hammer provides additional angles and masses and points of leverage for someone who doesn't have the necessary power to make the blunt poll of the back of a woodcutting axe truly devastating. The claw on the back of this hatchet can also act as a hook to make the damage coming out almost as bad as the initial impact of the blade. In the initial inquest, the state claimed that there was exactly that sort of hooking damage done to Andrew's head.

Instead of leaving the hatchet likely used in the murder in the box, Officer Fleet secreted it away to the vegetable keep, where the Bordens kept their jarred goods and shingles. I kind of see the logic—Lizzie could have stolen it away in the confusion—but it's a stretch.

In any case, the abominable chain of custody didn't stop the hatchet from being considered in a court of law. The chemist witness testified in pretrial hearings that there was not a smear or a drop or a speck of blood on the head or the handle. This other hatchet (or other two hatchets) fall out of the story after the chemist pulled the "innocent looking hatchet" out of his suitcase and said it had no blood anywhere in at trial—and that the hair on it was a cow hair.[17]

Fleet hid the handled claw-hammer hatchet in the vegetable keep on Thursday, August 4, the afternoon of the murders, and took it into police custody around the same time as Abby's and Andrew's bodies. But it took until after the weekend to find the final hatchet, the one the prosecution hung its case on, the one without a handle. But like an axehead hung before the wedge and handle are dry, the hang did not hold.

The police were searching the house again that Monday morning, four days after the crime, tearing apart anything that hadn't already been torn apart twice: they removed bricks from the chimney, sifted through the toilet tank, "dismantled the woodpiles, shoveled through boxes of ash, [and] emptied out the coal bins."[18] They could not have been more thorough "unless the paper was torn from the walls and the carpet taken from the floor."[19] They were finding nothing.

Another officer named Medley (another great name) was poking through the cellar again that day. He went back to the box Bridget had shown them, which was about the size of a shoebox—fourteen inches wide, five inches long—and filled with the kind of junk many abandoned basement shoeboxes are filled with—nails, bolts, old tools. It was all pretty grimy with ash. Medley took it down to root around a bit, check they hadn't missed anything.

They had.

A three-and-a-half-inch-wide blade, with a notch for roofing purposes and a hammer on the back. Probably the Bordens bought it years ago and forgot about it, or figured it was too broken to use and thus bought the claw-hammer hatchet Fleet had hidden. Stuck into the eye at the top was a chunk of the handle. Its end was splintered, the broken wood fresh as if the fracture were recent.

Medley "sung out" to his boss, who came to see it immediately.[20] They rubbed at a little spot on the metal, because that's a great idea, but couldn't tell whether it was blood. Then Medley wrapped the little tool head in paper, trying not to disturb the sediment on the tool so it could be tested, and he headed for the marshal's office.

The brokenness of the hatchet's handle fired the imagination of the police and prosecution. In their mind, the missing handle was a cause for alarm. The fact that there wasn't any blood on any of the other axes meant that there was blood on the handle, the one part of the axe they couldn't find. They talked themselves into believing that the break of the hatchet handle looked fresh, that some kindling in the fire was really the handle wrapped in newspaper and that the griminess of the axe was not the same as the other grime but ash, freshly applied to make it look filthy after it was washed clean of blood.

The little hatchet head was an effective prop at first. Their medical examiner made plaster copies of Abby and Andrew Borden's actual skulls to demonstrate to the jury that the broken-handled axe was capable of fitting perfectly in one of the worst gashes upon Andrew's skull—and that a new axe of the same make, which was in "very general circulation," would not fit in the same way.[21] This was probably the most powerful physical evidence that they presented in the trial.

Unfortunately, it was undercut by the evidence they'd already given about the hatchet and the circumstances of its discovery.

Their argument that the broken-handled hatchet was covered in some other, fresher grime to cover up the fact that it had been recently washed clean of blood simply didn't hold up even in police testimony—some said the grime was fine, others that it was coarse—and by the time the hatchet head actually reached people who were qualified to testify upon the exact qualities of the grime, it was coated in grime all over; there was no distinguishing the grime on the splintered handle, which was supposed to have been fresh and thus less grimy. They couldn't even agree about how the hatchet had gotten from the cellar to the police offices: one person said butcher paper, another newspaper, and the two ways they demonstrated wrapping the hatchet couldn't have been more different. The left hand didn't know what the right hand was holding, which is especially bad when one of the hands is holding a hatchet, handleless or not.

The police's whole theory was that Lizzie killed two people with an axe, cleaned the axe and herself, and then broke off the handle to burn it; one officer claimed they saw a bundle of newsprint and sticks about ten inches long and two inches thick, the size of a hatchet handle, in the flames.

Now, the logic of this is wild. The axe couldn't have broken in the attempt because there is no blood on the broken part of the handle that remains. Handles are strong. The axe handle was so tightly attached that it would have taken an additional tool to break it off, additional time that the overloaded sequence of events cannot sustain. Even with a vise like the Bordens had in the barn, it would have taken a lot of strength to break off the handle as quickly as Lizzie would have needed to in order to leave enough time to change her costume and hide the bloody dress. But even assuming she was able to wash and destroy this tool and dispose of her garments in the (at most) half hour after Andrew's death, she couldn't have used this axe.

Because the handle was still there in the box as Lizzie stood trial.

One of the officers who observed the hullabaloo when the broken-handled hatchet head was found that Monday after the crime said that when they found the axe, they found the handle too. He said this very casually, not realizing that Fleet had lied about it and would be recalled to lie about it again.

But there was one more axe yet to reveal itself.

A boy named Potter discovered the final claw hammer hatchet on a neighboring roof while trying to get a lost baseball, of all the prosaic American things. It measured about three and a half inches, same as the broken-handled hatchet. A carpenter said that it might be his. The tool had clearly been tossed a bit about by the weather, but underneath there was still a hint of gilt—the same substance the medical examiner had found in Abby's ruined right ear.

Though the axe was found while the Borden trial was at its height of mania and made a "genuine sensation" in the media at the time, attention to it quickly faded. The idea that it might be relevant was roundly mocked; the Fall River *Daily Globe* "sarcastically suggested that a search of the roof might reveal" a note from a sick friend that Lizzie insisted Abby had been talking about on the morning of her murder.[22] Discovery of this axe failed to make any lasting impact until nearly a century later.

I imagine that the prosecution—well on their way to an acquittal—had been so burned by the broken-handled hatchet that they didn't want to broach the idea of the weapon again; besides, a hatchet on a nearby roof extends the idea of Lizzie's physical prowess—a concept already relied on heavily in the prosecution's logic—to suggest that she was capable of having thrown it.

Rather, this axe suggests what most Lizzie defenders would say it does: that someone random did it. They bought or stole a new hatchet just for this purpose, they hid in closets and slipped in and out of doors while people looked away for a second, and they escaped over the rooftops, dropping their weapon on their way.

Or maybe not. This case is a maddening puzzle, where sliding one piece into alignment spins another one exactly where you don't want it. With basically zero physical evidence, everything is circumstantial. Forensics is a deeply flawed field, but it is better to have physical evidence than not, and this case is missing a bloody axe.

I look at the tension of the household, the locked door, Lizzie's flimsy alibi, and my logical sense says "obviously yes this is her, no one could have slipped in and out like that." And yet I try to imagine how the chain of events proceeded that hot morning, how she came out with no blood anywhere on her while Andrew's blood was still flowing, the lack of a weapon—and it all falls apart again. There's no single likely scenario for me. Those fifteen or twenty minutes between Lizzie's last conversation with Bridget and the discovery of Andrew's body simply do not leave enough time.

For a moment I can see her planning it out meticulously, getting Abby at just the right second in her routine so that Andrew and Bridget didn't notice. But how could she plan it and leave the matter of her alibi so utterly sketchy? The project management of the murder is either pristine or nonexistent.

And then I think, so maybe she got lucky. Abby said something that set her off; fueled by some old resentment, she decided to attack Andrew too. But that's just so much luck. And it doesn't answer the question how, in this frenzy, she stayed so calm with Andrew and Bridget. How did she protect her dress and hair? How did she hide the weapon? I can accept one or even two Broadway-quick costume changes or a housedress burned later. But at issue is not just a quick change. There is also the matter of the weapon. If she committed the murder with the hoodoo hatchet, she not only had to snap off the handle but also take it to the basement and cover it with ash and lock the cellar door behind her; if she used some other weapon, she had to wash or hide it so well it would never turn up.

Okay, so maybe Bridget was in on it. When I looked up the adaptations and saw a movie in which Kristen Stewart and Chloë Sevigny play Bridget and Lizzie as lovers in a conspiracy of murder, I scoffed. But to me that's less crazy than Criminal Mastermind Lizzie or Incredibly Lucky Lizzie. So many of the issues with Lizzie's guilt come from Bridget's say-so: her calm conversation with Andrew, her bloodless appearance between the murders. I don't *really* think they were colluding lovers—Lizzie's disdain for "Maggie" (Bridget), comes through clearly in the court testimony. But I can imagine that mistreatment brought them into each other's confidence. If I squint.

More interesting to me is the fact, often taken for granted, that Lizzie insisted that she did not murder her father and stepmother. People are very willing to disregard someone's insistence on their own innocence, but one thing I've found studying murders from many different time periods is that people who are guilty quite often confess in one way or another. Of course, some people have the tremendous capacity to lie, and some confessions are coerced: I'm not saying that people who confess are *always* guilty or that people who insist they are innocent are *never* guilty. It's just that people who have killed often find it hard to pretend they haven't. Lizzie's insistence on her innocence isn't any kind of proof either way, but to me it's an important data point.

I can imagine other suspects: an assassin hired by Emma or Lizzie, one of their neighbors. There are a couple of random suspects who don't strike me as especially likely: there was a buggy right near the Borden house around 10:50, and about two weeks before the murder a man came to the Bordens' door and confronted Andrew. Or maybe it was a serial killer. Very few murders are the work of serial killers, but some of them are.

The reason that Emma Borden thought Lizzie was innocent was not because of her character. It wasn't the younger woman's honesty, independence, or compassion toward animals that made Emma believe she could never do such a thing. It wasn't Lizzie's alibi, which is not great. It wasn't that her clothes were free of blood, but that something else that was missing.

The hatchet.

The axe of easily the most famous American axe murder, the axe that the Lizzie of the public imagination used to take her eighty-one whacks, the hatchet that the journal of Lizzie Borden studies is named for.

If Lizzie did it, she didn't use an axe. Maybe a cleaver, which is sometimes called a meat-axe. But definitely not the broken-handled hatchet.

"The authorities never found the axe or whatever implement it was that figured in the killing," Emma said in her only interview, given in 1913. "Lizzie, if she had done that deed, could never have hidden the instrument of death so that the police could not find it. Why, there was no hiding place in the old house that would serve for effectual concealment. Neither did she have the time."[23]

Fitting for the coldness of the Borden demeanor, there is no emotion to this argument. Because it's not the passion that makes this murder toll as strong and clear as the Fall River city-hall clock throughout American folklore. It's the puzzle: the locked doors, the two hours between the murders, the twenty minutes between Andrew sitting down to a nap and the discovery of his body, Lizzie's shifting position in the house. The axes, shuffled and tested and studied for centuries, slotted into one theory after another like a children's shape-fitting game. All found wanting, none satisfying the bloody riddle.

Shingling Hatchet

1914, Spring Green, Wisconsin

The axe has a profound relationship to architecture. It clears the land and processes the wood that creates walls, floors, roofs, furniture, heating. Not every culture uses wood for housing, but in most building traditions wood is a major element in taking permanent shelter and making a comfortable home.

In 1914 axes were still an important tool in making comfortable homes. Logging was still mostly a one-on-one affair: one lumberjack, one axe, one tree. Chainsaws existed, but they were still a few years from widespread use. But industrialization was creeping in. The process was getting much more automatic—the lumber coming to the worksite already cut, the concrete premixed.

In the early twentieth century, people still had a hand in the building and repairing of their own home. As the shingles started to come precut, as the tools got more complex, the skills involved were harder to acquire and easier to sell. The roofing hatchets in the Borden basement were picked up by a cheap man who wanted to save on his repair bills—simple tools, with a claw for pulling out nails and a notch for measuring. As the task became more specialized, more and more parts were added to the cheek of the axe-head: magnets, blades for cutting, gauges. A tool for professionals, still common but less necessary than ever.

TEN

A FULLER MEASURE OF LIFE AND TRUTH, AT ANY COST

In 1901, Frank Lloyd Wright was eager to move on from the axe. Nothing against the axe specifically—it's just that he was in favor of machines with more than one moving part.

Only thirty-four, Wright was already a national figure in architecture. He was at the vanguard of the arts and crafts movement, which advocated for a return to the principles of nature in art and design. Mission-style simplicity took priority over Victorian ornamentation; handmade objects were preferred to industrialized products.

Wright's take in his 1901 speech to Hull House in Chicago was that machines, used well, could better reveal the properties of natural resources like wood. In his view, materials like concrete, glass, and steel were natural themselves. "Tools to-day are processes and machines where they were once a hammer and a gouge," he said.[1] Just as *Homo heidelbergensis* in their caves could not reject the hand axe, Louis Sullivan in his skyscrapers could not reject the Bessemer method. If the machine could make greater beauty for a greater number of people, then Wright would happily walk into the steel forest without a thought of the hammer, gauge, or axe.

Wright dissented from some of his arts and crafts peers with his enthusiasm for steel furnaces and cement mixers. But he was all in on the overall philosophy and its approach to architecture, especially its focus on domesticity: if mass production created shoddy furniture and homes, that was liable to damage the family itself. One of Wright's lasting contributions to American architecture were the open plans that drew the family from small bedrooms and liminal foyers and hallways into large collective gathering spaces by combining dining and living areas. Home and family were a religion for Wright in 1901, the foundation of his burgeoning ambitions. The pursuit of freer love cut through his life like a chainsaw.

For the first twenty years of his career, Wright projected an image of domestic bliss with his wife, Catherine, and their six children in the tony Oak Park community outside Chicago. He grew up as the obsessive favorite of his mother, Anna, who had objected to his marriage to Catherine in 1889; "Kitty" was young, beautiful, and from a wealthier family, and Anna was furious about it. By the turn of the twentieth century, Kitty had figured out how to deal with

Anna and had found domestic bliss. She and Frank had six happy and indulged children, and Kitty had a bustling civic and social life.

But by 1904, Wright was starting to chafe at the expectations and responsibilities of family life. Wright wasn't a man who liked the stability of a balanced checkbook and a steady marriage; he wanted excitement and glamour and to go into serious debt by buying three concert grand pianos and a dozen Chinese rugs in one shopping spree. The girl who had seemed ideal in 1889 had "made the fatal error of becoming a matron," as William Drennan put it in his excellent *Death in a Prairie House*.[2] Wright stopped going to church and grew his hair out.

Architecture was one of the few places where wives were expected to make important decisions; much of Wright's major early work was first covered by women's magazines of the era. He was used to meeting women who threw themselves into the lives of the home. Edwin Cheney's wife was different.

Mamah Borthwick Cheney had always dreamed of a life beyond the home. Born the daughter of a machinist in Iowa, she grew up in Oak Park before studying literature and languages at the University of Michigan. This was a school that had been coeducational for more than twenty years; one of Mamah's contemporaries described it as her "first taste of emancipation."[3] She got her master's degree and worked as an educator and librarian before she finally relented and agreed to marry Edwin Cheney, whom she knew from the University of Michigan, in 1899.

She entered into domesticity with reluctance. Marriage gave Mamah security and a position in the world; she could experience sex and love without becoming a social outcast. But domesticity was a rough fit for a woman who valued herself less for her sewing or cooking or decorating than for her translating and writing and creativity. "For a middle or upper-class woman, getting married meant being sentenced to the purdah-like realm of the suburb," wrote Anne D. Nissen in her doctoral dissertation on Mamah. It's a familiar story, Nissen wrote in 1988—when a woman becomes sexual, she also becomes a threat. Motherhood and homemaking are supposed to neuter that threat. Mamah made some effort toward fulfilling the domestic ideal, meeting Kitty Wright and Ernest Hemingway's mother, Grace, in a women's group. She even befriended Kitty, which is probably how Frank was commissioned to design a home for Edwin and Mamah Cheney in 1903.

The architect was immediately infatuated with Edwin's urbane and well-educated wife. Mamah was a sharp contrast to Kitty, who, in the words of her husband, "wanted children, loved children, and understood children. She had her life in them."[4] Mamah never had such a focus on her children, John and Martha. During her marriage, she relied heavily on domestic help to take care of home and children while she took classes at the University of Chicago. This was sexy to Frank: "one of the things that attracted him most to Mamah was her relative indifference" toward being a mother.[5]

In June 1909, Mamah left for Colorado with the kids,

telling Edwin that she wanted to visit a pregnant friend; in October, she told Edwin that the friend had died in childbirth and that she needed him to come pick up the kids. She was gone by the time he got to Colorado.

That same month, Frank was also up to something. Frank borrowed a ton of money from several different people, literally gave away his thriving architecture practice, and told his thirteen-year-old son that he was responsible for a nine-hundred-dollar grocery bill as the man of the house. His position was that if society wanted him not to abandon his wife and kids, then that was society's problem.

The lovers ran off to Berlin. They were caught registered as man and wife at a Berlin hotel by a tabloid reporter, ruining Frank's reputation for years. But the European trip was creatively productive and harmonious; he published a portfolio of architectural renderings in Germany known as the Wasmuth portfolio (after the publisher) and worked on new designs in Italy, and she worked as a teacher and translator in Berlin. The Wasmuth portfolio was rapturously received; though Wright's ability to generate paying work was limited by his recent transgressions, his renown was on the rise once more.

The couple were in the glow of new love and creatively fulfilled; it was an idyll out of the nineteenth-century Romantic poetry Wright loved. The relationship was all the more valuable because it couldn't last. In a weird letter to his mom, Frank decried that their return would be a victory for "THE INSTITUTIONS," calling himself a "weak son who was infatuated sexually" and who would likely be forced to abandon the source of his infatuation to the winds of public shame and, potentially, to a return to her unhappy marriage: "a craven return to another man, his prostitute for a roof and a bed, and a chance to lose her life and her children."[6]

Frank wasn't used to considering other people, but he knew that Mamah had even more to lose than him in terms of status: "an outcast to be shunned."[7] He seemed to grasp some of the basics of Mamah's feminism: that men don't own women, that women don't own men. But I suspect Frank didn't care much about the philosophy beyond using it as a rationale to leave his wife.

The independence he admired in this "brave and lovely woman"[8] made her a monster to the rest of their world in Oak Park. Catherine wasn't the only one who felt her to be "a force against which we have had to contend." Her former friend, her husband's new lover—she wasn't a woman but "a vampire."[9]

This kind of social reaction is exactly what drove Mamah to her mentor, Swedish feminist Ellen Key. Key's "new morality" drafted a new vision of the family unit: she advocated for birth control, access to divorce, breaking down class systems, and marriage based on love and partnership. But her work wasn't all progressivism: she was as fixated as any suburbanite on the all-important role of the mother in domestic life. And, like a lot of white feminists from all waves, Key could be racist, her logic often wrapped up in eugenics.

"Free love" is the enduring message of Ellen Key's work, and these were the ideals upon which Taliesin was founded: "the basis of this whole struggle was the desire for a fuller measure of life and truth at any cost."[10]

Truth was a fixation for Frank, perhaps because it was an ideal that could be manipulated more easily than his budgets. Frank's abandonment of his wife and architectural practice was hurting his career and his ability to take out more loans. Kitty bought his spiel about radical honesty and they reconciled very briefly. He got another loan based on his restored reputation, and then he left again, off to Taliesin, his "architectural self-portrait."[11] With Mamah, of course.

Edwin Cheney had an admirable lack of vindictiveness. He was amiable by nature, beloved by the community in a way Mamah never was even before the divorce. He knew from their courtship that he was never Mamah's first choice; I think, like Frank, he admired and was attracted to her independence and intelligence. Losing Mamah as a wife was a blow, but one he could withstand; he was remarried by 1914. He held primary custody of their kids, but he did not fight her time with her children and, unlike Catherine, made no restrictions about their spending time with Frank.

Taliesin was built on Wright's uncle's farm, purchased by Wright's mother, Anna, in 1907; the name is a reference to his Welsh roots, meaning "shining brow." It is set against a rolling hillside carved by an ancient ravine, unique because it was never covered by a glacier. He intended it to be more than just a fortress for retreat with Mamah: a place of education, performance, and sustenance. It was to be a home, and school, and theater, and fully functional farm, with its own icehouse and granary and power supply. Frank was making himself his own little dollhouse, where he was the accepted authority in matters professional, creative, and domestic.

But while Mamah and his students and workers were willing to live in accordance with his values and principles, the people of Spring Green were not thrilled about him and Mamah playfully fording streams while clad in lingerie.

Ostracism of Frank and Mamah extended to her children too. There were a few children in the area surrounding Taliesin, but many of them were not allowed to play with the Cheneys, especially not at the house. Edna Kritz was an exception. Her father grew up with Wright, and though he didn't approve of how he was living now, he felt that the children were innocent. Edna much admired Martha especially, her pageboy haircut and her sapphire ring, and they often played with dolls together at the "Love-Bungalow," as the papers called it, or the "Love-Castle."

Wright summoned one of the newspaper reporters making his life difficult to hear his side of the story. A tactic that always works. He brought them up to Taliesin on Christmas, both Frank and Mamah in robes, sitting next to a roaring fire in the limestone hearth that feels as if it was carved out of the bedrock of the land. He wanted to explain a few things to the newspaper reporter: the failure of his marriage to Kitty was the fault of his parents; he hated being called Papa and his buildings were his real children; Kitty never understood him; he was not subject to the customs of the average man; and Mamah agrees with everything he said.

As a mentor, teacher, employer, lover, Wright dominated his acolytes. In his later years, Frank's narcissism was treated as an eccentricity, part of the price of admission for knowing the great man. I wonder how Mamah—who had left marriage in part because of its constriction upon her freedoms—coped with a relationship where her voice was expected not to rise except in harmony with his. Even her own work could not resist the force of him: her translation of Ellen Key's work also credited him, so vociferously did he claim her things as his.

His architecture is famed for its sense of flow: compression in liminal spaces like hallways and thresholds squeezing you toward the release of a spacious living room or a spectacular vista. But he created a tense professional atmosphere. Everything was based on the great man's capricious whims.

Julian Carlton hated Taliesin almost as soon as he arrived in June 1914. Before the summer was over, he would be dead, and seven other people at his hand.

Spring Green was and is a tiny town of less than a thousand people—1,566 right now, less than half that in 1914—who are about 99 percent white. Carlton and his wife, Gertrude, came from Chicago, a diverse, bustling metropolis. The job he'd had there as a janitor wasn't ideal, but leaving it behind wasn't worth the tradeoff of having to live at Taliesin. The Carltons were deeply isolated by the move from a huge city to a small town, as was Mamah. The other workers could go to a tavern and blow off steam, but that wasn't a place where a young black couple would feel comfortable. Carlton didn't drink anyway, and he didn't want to see his colleagues socially.

The Carltons were experienced in service—Julian had worked as a Pullman porter, a prestigious position—but the other people who worked there weren't servants. The couple came to Taliesin by way of a recommendation from a caterer named Vogelsang, after the Carltons worked for his father. They had been married about two years. Carlton was in his early thirties, possessed of a mild manner but prone to sudden eruptions of mental illness.

There were five other men working at Taliesin that summer. Draftsman Herbert Fritz was barely out of his teens. Sixty-six-year-old Tom Brunker was a handyman and foreman and the father to ten kids. Middle-aged Swede David Lindblom was a landscape gardener. Billy Weston, thirty-five, had built most of Taliesin; he was Wright's favorite carpenter. He had his thirteen-year-old son Ernest with him.

Emil Brodelle was thirty and a man on the rise. His aerial drawings had earned him the esteem of the man they were all there to please, and his personal life was looking bright too: he'd recently become engaged to the sister of a close friend. Perhaps all of this good fortune gave him a bit of an ego. We've seen a lot of prideful men and women taking out their frustrations after personal struggles, but sometimes success can have the same effect.

In any case, Brodelle was a racist jackass to Carlton. On August 12, 1914, two days before the murder, Brodelle ordered Carlton to saddle his horse, and Carlton took offense. He was a servant to the house, to Wright and Borthwick, not to this guy. Apparently, Brodelle took offense to this offense taking, and he called Carlton a "black son of a bitch."[12] It seems unlikely that this was the only racist thing said to Julian or Gertrude, nor can I believe that Brodelle was the only man saying reprehensible things to the couple.

Carlton raised the issue with Wright, later saying that everyone was "picking on him."[13] The Carltons' relationship with Wright and Borthwick is one of the parts of this story that changes the most from one telling to the next. Drennan insists that Frank and Mamah were extremely impressed with the Carltons—immediately after the murder, Wright called Julian and Gertrude "the best servants I have ever seen" and commented that Carlton had behaved in a totally normal way earlier that week. Newspapers at the time suggested that their frequent target Mamah distinctly disliked Julian and had, in fact, fired the Carltons shortly before the murder. Either way, they were planning on returning to Chicago that day; Wright was advertising for additional domestic workers in the paper that week.

There were other signs of rising tension. Weston described Carlton as "hotheaded," and Lindblom mentioned a confrontation in which Carlton said, "if anyone around there did him any dirt he would send them to hell in a minute."[14]

Julian was sleeping with a hatchet in a bag by his bed in

the days before the murders, according to Gertrude. He was waking up at night, searching in the darkness, talking about people being out to get him. The hatchet he used was not flared or bell-shaped. It was a rectangle. Four right angles, two sides equal in length and parallel with each other. The head jutted out from the handle in a way that reminds me of the cantilevered balconies and terraces that distinguish the work of Frank Lloyd Wright.

Wright's projecting roofs notoriously require a lot of maintenance. At his most famous property, Fallingwater, this fact is part of the tour. Guides not only don't try to hide the leaks, they'll point to the cheap plastic containers they use to contain them. Wright worked with the natural features of the land rather than armoring his houses against the elements; putting waterfalls in residential homes is not great for the foundation or the roof. The jutting angles of the roofs cause a lot of stress that allows the rain in, reminding everyone that nature will always have the ultimate advantage. Taliesin had multiple buildings—the Hillside School where his aunts taught, his sister's residence Tan-y-Deri, the Romeo and Juliet Windmill— and shingles at Taliesin were different for different houses. Clay for the hillside buildings and cedar for the house itself, the kind that requires a shingling hatchet. And Wright was always pushing the boundaries of concrete and wood and steel; not all of his experiments worked, and so repairs were a constant feature of life at Taliesin.

Odd jobs were a part of Carlton's position at Taliesin in the summer of 1914. He did work on the house itself sometimes, but cooking and serving Wright and everyone else at Taliesin was a major part of his work; by some accounts, he was a butler. The Taliesin website describes him simply as an "angry Taliesin domestic employee."

John and Martha Cheney didn't want to be at Taliesin either. This was not twelve-year-old John and eight-year-old Martha's first trip to Taliesin: they came for a month every summer. The novelty and splendor of farm life and Wright's scary bear rug had long since worn off; rolling hills are no substitute for actual amusement, and a fine collection of Japanese prints were not a great substitute for companionship. Perhaps the unhappy children added to the tension of the household in the days leading up to August 15.

In some versions of the story, Mamah sent a last, desperate telegram to Frank, who was working in Chicago, before lunch on August 15. "Something terrible has happened," she purportedly wrote.[15] An awful last sentence for a woman who found power and purpose in the written word, who had many chapters more to live. But I don't believe it. Like Catherine Howard's dying declaration that she would have rather died Thomas Culpeper's wife, it's too narratively neat. Unless there was some other terrible event that was lost to time, the detail makes no sense.

Julian served lunch to his victims first, in his immaculate white jacket. Gertrude made soup for their last meal at Taliesin.

As Mamah and her children, Martha and John, sat down to eat, Carlton raised the shingling hatchet. They were on a screened terrace looking out to the view of the rolling hills. Carlton crushed Mamah's skull, and she fell to the floor.

John went next, in one blow.

Martha tried to escape. She ran from the terrace to the sunken square garden at the entryway to the house. Technically it's called a loggia—one of those Italian flourishes Wright delighted in after his idyllic 1910 European trip with Mamah. Mamah planted flowers in the loggia; in August they were in full bloom.

Carlton caught Martha and hit her repeatedly. He left her alive on the ground. Her playmate Edna was one of the first people to arrive after the fire Carlton would later set; she was already on the way to see if Martha wanted to ride horses. She found Martha on the loggia still, her clothes burned off but the girl alive. Edna became a poet later, a regionally successful one; she remembers screaming "that's not Martha!" inside her head, until she saw the sapphire ring.[16]

Gertrude was afraid of her husband and knew that he was not in his right mind; he could turn on her in a second. She grabbed her "best hat" and escaped through a basement window, running to the road nearby.[17]

After this, Carlton got the remaining soup and served the workmen. They were about eighty feet or so away from Mamah and her children, in a cramped sitting room down the hall from the kitchen; the narrow hallways and closed door dampened the screams. Either Carlton changed his jacket or they didn't notice the blood. "It is unlikely that they were used to taking notice of the black servant, anyway," noted one study of the murder.[18] Carlton locked the door behind him when he exited the room and began quietly dumping gasoline on the floor.

The young draftsman, Herb Fritz, survived the event; he later said he had just noticed the liquid spilling into the room when Carlton lit his pipe and dropped the match in the gas. The ground shook. The man, frenzied with terror and fire, tried desperately to break down the door. Fritz found a low window and threw his body through it, down toward a creek at the bottom of the hill. Outside the house, Carlton waited with his hatchet.

Emil Brodelle—the man who had insulted Carlton two days before—started to get out the same way Herb had, but Carlton was upon him quickly, driving the hatchet into his head at the hairline. On the other side of the house, the remaining men were spilling out into the loggia; Carlton used the back of the axe ineffectively on Billy Weston, but killed his thirteen-year-old son Ernest in one blow. David Lindblom was hit in the back of the head. Tom Brunker's head was almost as destroyed as Mamah's.

Several people on the loggia were still alive. Lindblom and Billy Weston were able to make it to a neighbor's house a half mile away to call for help, and then Weston dragged himself back to Taliesin to try and stop the fire. When Carlton thought he had hit everyone, he returned indoors to douse the bodies of Mamah and John with more gasoline.

Still holding the axe, he crept down into the unlit furnace in the basement and swallowed muriatic acid, purchased the previous week. While Billy Weston and David Lindblom tried to get help, Herbert Fritz hung on to life with all he had, and Martha Cheney endured the last few hours of her eight-year-old life, Carlton hoped that death would come sooner for him. It didn't.

In the weeks between the murders and Carlton's death from starvation in a prison cell, he discussed the case very little. He never gave any indication as to his motive or frame of mind beyond his anger at Brodelle.

Carlton chose a day when Wright wasn't there to set his architectural self-portrait ablaze. He spared the architect, but still wanted to see his dream burn. Wright was in Chicago, attending to Midway Gardens entertainment park he had recently opened in the Hyde Park neighborhood. Within two years it was bankrupt, sold to a brewery who "cheapened" Wright's aesthetics to "suit a hearty bourgeois taste."[19] By 1930 it was destroyed, another doomed dream.

Taliesin would be rebuilt—and burn—and be rebuilt again, then foreclosed upon, and then regained. Wright would finally divorce Catherine, marry and divorce and remarry, have another child, design thousands of buildings from his Taliesin studio, with the students he taught there. When he died and was buried in 1959, it was next to Mamah, whose grave he could not bear to mark during his lifetime—and then he was exhumed and buried again in Arizona. The peaceful harmony with nature Wright sought to express is haunted still by the natural pressures of water, fire, and earth, and the human habits of bigotry and violence.

Boy's Axe

1949, Los Angeles, California

In the years after Taliesin, "axe murder" was a staple of newspaper headlines. The term had its peak in the late 1920s and through the 1930s. It was still in use by newspapers in 1949—Jake Bird, a serial killer in the Pacific Northwest, had been described frequently as an axe murderer when he was caught the previous year. But the shock had worn off of axe murder, and it had begun to cross over into farce. In the 1940 comedy *I Love You Again*, Myrna Loy says, "I don't care if he was an axe murderer!" of her lover William Powell (not playing *Thin Man*'s Nick and Nora this time). Noir writer Raymond Chandler made a bit of a dark joke in a 1945 letter about Taki, the cat he brought on his travels. If Taki didn't like a place, "chances are there was an axe murder there once and we're much better somewhere else."[1]

By the Second World War, the axe seemed silly in comparison to the military's mechanized violence. Technology was moving so quickly forward. Many axes had been melted for the war effort, and they were not all getting replaced. After the horrors of war and the Great Depression, the nation's capacity for shock was low. Innocence was at a premium.

A boy's axe is a smaller axe, light enough to use with one hand but with a long enough handle for two hands. It's not an essential tool for survival so much as an accessory for camping. Something to keep around the house to cut wood with when you feel like it. Handy enough to reach for but not enough to get most jobs done.

ELEVEN

"WHOEVER COMES OVER, I GIVE ANYBODY CANDY"

Linda Joyce Glucoft's last words to her mother were "I'm going to go play with Rochelle."[2] Linda was a quiet but happy and popular six-year-old girl with light brown curls like Shirley Temple. On the day she died, she was wearing red shoes with yellow socks and a blue plaid dress; her aunt would later tell the newspapers that she loved plaid.

November 14, 1949, had been a totally normal Monday. At 2:30 p.m., her mother, Lillian, picked up Linda and her eight-year-old brother, Richard, from school. At home, Richard changed, and they went back out to take him to Cub Scouts. Then Lillian and Linda went and did some "household errands."[3] They went home, where Linda had some orange juice and cookies, and then she went to play.

Lillian knew that Rochelle's grandfather, Fred Stroble, had "played with"—molested—Linda.[4] But she wasn't concerned. She might not have been aware that he was living there again, but even if he was, he was a neighbor, not a stranger. He'd been nice to Linda, given her candy. Lillian thought he was harmless.

He wasn't.

Lillian wasn't a bad mom. She cared about her daughter's health and well-being. It's just that child sexual abuse was not really something you made a big deal about in 1949. It was a misdemeanor. And if you pursued it, there could be a lot of severe social consequences—especially for Linda herself.

If Lillian tried to ensure that Stroble saw consequences for his behavior, the most likely outcome would be that Linda got a reputation as a liar. Lillian could have hoped that Linda misunderstood what was happening, or even that she made it up—she was taught that she might not be able to believe her child. She was following the established best practices for mothers whose children were sexually assaulted.

This seems wild. Especially when you take into consideration that Linda had already been molested and kidnapped on a separate occasion when she was three, on a golf course. This detail is buried in the coverage, and it startled me when I came across it. This poor kid lived the life that stands behind the fears of every helicopter parent. But in the mid-1940s, as long as children emerged from abuse alive, the discussion was often closed. By ignoring it, Lillian was doing what was thought to be best for her daughter.

The threat of violence is always an element of communal

life, and it's not generally a reason for children to forgo playing with their friends. Though I think most conscientious parents today would not have let their daughters go back to the house where someone had molested them was staying, Lillian was doing what conscientious parents did at the time. She told her daughter to say no and leave the next time, and not make a fuss.

Besides, she thought, even if Fred was around, Rochelle and her mother, Sylvia, would be at the house as well, buffers between him and Linda. She was wrong there too. Just ten minutes before Linda's arrival, Rochelle and Sylvia left to go to a classmate's birthday party.

Fred was left alone to receive Linda, with his candy, and his ice pick, and his axe.

Fred Stroble was evading bail for sexually abusing other young girls when he killed Linda. Just that April, he'd been charged with "inciting the passions of a child," a ten-year-old girl, in nearby Highland Park. He was already a "known offender," but the charges were just misdemeanors.[5]

Stroble was a small person, five foot five, and slender. He immigrated from Austria to America as a teenager in 1901. He met his wife on a cruise from New York to San Francisco and married her in Panama; he found a career as a baker and had at least one child. He was hit by a streetcar at some point, after which he was prone to dizzy spells.

On a 1941 trip to Honolulu with his wife, Stroble was caught in the act of sexually assaulting a nine-year-old by the child's teacher, but there were no apparent consequences. In the same visit, he witnessed the attack on Pearl Harbor. In 1943 his wife had an (understandable) mental breakdown that he said turned violent. She was committed to a mental hospital. In 1946 he lost his job as a chief baker at a large bakery and had "too much time on his hands."[6]

"Since then, I'm lost," he said in his confession. "Since then I'm not good. I have too much freedom and I don't know what to do with myself; got in to all kinds of mischief."[7] Neither his drinking nor his predatory behavior were new, and now they were set to escalate.

Linda was at the end of a very long line of young girls who were abused by Fred; he was able to give the name and address of four or five—meaning that he knew who they were, so perhaps they were friends of Rochelle or his grandson, Frederick. He also related that he would pursue crowds of children for his own gratification.

Linda could not have been the first to resist, and so I do wonder if there were others he killed to keep quiet. While the legal consequences of being caught molesting children were relatively mild and there were a lot of people who were perfectly willing to look the other way, child sexual abuse was not actually smiled upon—it was just not talked about. And the country's long tradition of politely ignoring dirty old men was drawing to a close. Perhaps this is part of why he decided to escalate his behavior and murder Linda; as the stigma of the act grew, so did the consequences Stroble looked to avoid. He was already facing down legal trouble; being caught again would make things much worse.

Sylvia professed to be close to her father. Her husband, Ruben, said he knew about "the incident on which he was arrested" a few months before but kept it from Sylvia to protect her.[8] They later testified against him, so I believe they regretted their trust in him. Fred lived with Sylvia and her family up until the day of his court appearance for his offense against the Highland Park girl, when he absconded to Mexico. He stayed there for months and returned a week or less before the murder, giving the chilling reason that he missed his grandchildren. In Ruben's telling, he had informed Stroble the day before the murder that he needed to turn himself in or leave their home forever.

Stroble had known Linda for about two years before he murdered her. Linda's father, Jules, commented in the aftermath that Stroble would often say, "I like Linda."[9] He gave her a chocolate bar every time he saw her. "Whoever comes over, I give anybody candy."[10]

On the morning of November 14, Stroble got up at 5:30 a.m. and shaved his face. He had a cup of coffee and then walked around until about 2 p.m., drinking wine. I suppose LA was walkable back then. When he came back, he switched to whiskey; his daughter and son-in-law did not notice his drunkenness. He was alone when Linda came asking to play with Rochelle.

The candy bar in his pocket was melted, but he enticed Linda by promising her some chocolate that he had in the icebox. Then he led her to the rear bedroom. She protested, said that she wanted to play outside, but he threw her on the bed. "I never was in a state like that in my life," he later said.[11]

Linda fought. She screamed and wrestled to get away.

The details get very gruesome here. He held her down, told her not to "holler."[12] When that didn't work, he strangled her. First with his hands; she quieted at first, but then began to weakly squirm. He took a necktie and knotted it around her neck. When he noticed more movement, he got a ball-peen hammer; he covered Linda's head with a brightly colored blanket and hit her on the temple. Then he dragged her—he was too weak to pick up her six-year-old body—to the kitchen next to the incinerator.

She moved again, so he got an ice pick; he pierced her chest with it, and then twice more through the back so that he knew he got the heart.

And then he went to the garage for the axe.

This was an urban area, central LA. In the previous years, many axes had been melted down to make guns and tanks for the war effort. But the axe was still something any homeowner would have handy for gardening or breaking down firewood. The axe he used was not a hatchet or a maul or a standard two-and-a-half-pound axe. From the photographs outside the home, where an officer holds the axe lightly, I think it was probably a boy's axe—slightly shorter and thinner than a full-size axe, appropriate for minor woodcutting and camping. Fred was a small man more used to baking bread than cutting wood.

He made a real mess of her small body. He hit her in the head, and then her body as many as a half-dozen times. He said later that his intent was to make sure she didn't suffer any further. "I don't like to see anybody suffer, not even a dog or a cat. I said, 'Well, the damage is done, I might as well finish the job.'"[13]

Somehow, he wasn't done. The long knife was the fifth weapon he used after his initial attack, each to make sure she was dead. He stuck it in her neck from the back like he'd seen at the bull races.

After that, he dropped the knife by her body and went inside to turn off the fire under the potatoes that had been cooking all the while. He covered her body with some cardboard boxes and stuffed her underwear in the incinerator. Then he put on a coat before he caught a streetcar on Venice Boulevard to Ocean Park in Santa Monica. "I think I go out to Ocean Park and have a last night good time and jump over."[14] It makes me shudder to think of what this man would consider as a good celebration for his last night on earth, but whatever he did that night he decided not to jump after all.

Lillian got worried an hour and a half after Linda left. Linda's brother had come home from Scouts, changed, gone out to play, and come back in with no sign of his sister. Dinner was ready.

At 5:10 they went over to Rochelle's house. Of course, no one was home.

At 5:15 they raised the alarm around the neighborhood, knocking on doors. It is difficult to imagine Lillian's anguish at this moment, how frantic she must have been, picturing the worst and hoping it wasn't real. At 5:45 they called the police; at 6:00 Sylvia and her family came home.

By 7:00, there was a search party of twenty-plus people performing a "methodic" search of the neighborhood.[15] At 7:30 Linda's father, Jules Glucoft, a commercial artist, came home to a nightmare. Around 8, Sylvia's husband, Ruben, remembered what his father-in-law had been arrested for; he informed the police and gave them a picture, sharing that Stroble might be in Mexico.

The search wore on into the early hours of the morning; by 5 a.m. there were scores of policemen out on the search for Linda and Stroble.

At 7:12 a.m. Linda's body was found in the kitchen of her friend Rochelle's house.

The reporting of this part of the awful discovery was quite intimate. Reporters and photographers were there in the Glucoft home when Lillian and Jules and their extended family were informed of her death, taking pictures as they collapsed and wailed. They photographed Stroble's family too. Rochelle and Frederick were shown being taken away from the scene—the caption took pains to note that they were not informed of what was going on. And there was a shot of Sylvia and Ruben too, in obvious agony. They recorded Linda's brother coming home, crying, "Mother! My little sister is dead!" and how their mother cried for her baby.[16] They photographed not only the grieving mother in unimaginable pain, but the body itself, under a Maxwell coffee cardboard box, next to the axe.

Modern true crime, for all its intrusiveness and exploitation, is no longer privy to moments like these. Crime scenes had very porous boundaries for a long time. Like the idea that child molesters were not benign, the idea that crime scenes should be preserved had only started to be voiced about fifty years before Linda's murder; things like police tape wouldn't come until the 1960s. The media at the time had significant access to the locations where bodies lay and where survivors were informed but no longer has the strength in numbers to muster such aggressive coverage.

But with that force, they were prepared to deliver a sense of righteous anger and desire for bloody justice to their readers. Fred Stroble's name and photograph from his previous arrest were everywhere.

Despite being only one of five different weapons, the axe was the focal point for descriptions of the process of Linda's killing. Likely this has something to do with the photographs, which heavily featured the axe, both at the scene where Linda's body was discovered and as it was carried by detectives. But the axe also served to underline the brutality of the attack, which was especially potent when contrasted with Linda's innocence.

The newspaper stories initially conveyed shock and outrage with a prurient undertone, which is a profitable combination. A lot of time was spent on the details of the appearances of anyone who was female: Linda's chubby cuteness is mentioned constantly, Sylvia's outfits as she testified against her father were closely detailed, and Lillian is called "the pretty mother" in the same sentence that describes how she "shrieked incoherently" when she learned of her daughter's death.[17] The "sex-mad killer" aspect of the case was more prominent than the axe murder side of it to newspapers.[18]

Though photos of Linda, the crime scene, and the grieving Glucofts dominated early coverage, the focus over the long term of the heavily covered case was Stroble. He was a "symbol of unalloyed evil."[19] The suspense of his disappearance and the subsequent manhunt provided the momentum necessary to create widespread interest in the story. He was thought to have escaped to Mexico for a time, but he never left the city. His name was everywhere, his face above the fold, and he didn't leave Los Angeles, so within thirty-six hours of the discovery of Linda's body he was seen having a beer in a café and apprehended. He went quietly and confessed fully, saying that he didn't deserve to live; in court, he kept his eyes from "Linda's red shoes and an axe which lay on the counsel table."[20]

Stroble's submission to authority did nothing to abate the intense sentiment against him. He claimed senility and alcoholic dementia at his trial in January, which did not go over well; it took fourteen minutes for the jury to return a verdict of guilty. The brevity of the jury's deliberation and whether it came as a result of their prejudice from the extensive coverage was the subject of Stroble's appeals, which went all the way to the Supreme Court. But the appeal failed, and he was executed in the gas chamber on July 26, 1952.

Even as newspapers exploited their readers' fear for profit, some coverage reflected the more forgiving attitude that had allowed Stroble to get away with raping children for decades.

Dr. Marcus Crahan, who examined Stroble for the district attorney, gave an explanation of Stroble's motivations that was by turns generous and deploring. Stroble was a lonely and forlorn person who loved and missed Linda, it began. But by the end he described Stroble as one of the "idle old men" who exemplify the "true sex psychopath."[21]

The sex psychopath was one of the prime boogeymen of the day, like human traffickers in Walmart parking lots today. After World War II there was a growing awareness that child molestation is not a minor offense, and that realization led to a guilty overcorrection: predatory perverts were believed to be around every corner. This made it easier to believe that kids were still lying about the people who, we now know, were most likely to abuse them, like priests, teachers, or parents. Fred Stroble was a validation of the righteousness of this fear-mongering.

Dr. Crahan had another theory about Stroble: "Possibly he was a homosexual and didn't know it."[22]

In response to Stroble and a perceived wave of sexually motivated murders of children, Governor Earl Warren called a special session of the California state senate to address it. From this session resulted several bills. Some of them were necessary, reclassifying the rape of children as a felony. But one of them directed the Department of Mental Hygiene to look into how to cure sexual perversions. Like homosexuality.

The law didn't assign any punishments, but the idea that being gay was something to be cured remained on the books until 2010, when Linda's younger sister campaigned for it to be repealed. As social norms crawled in the right direction by taking child sex abuse seriously, they strode in the wrong direction by further restricting queer rights.

Linda's murder inspired the filmmaker Sid Davis to find his calling in social guidance films—those short, cheesy films shown to children by schools that tell them how to behave. (Another example of a social guidance film is *Reefer Madness*, which was also inspired in part by an axe murder—Victor Licata's 1933 massacre of his family in Tampa.) *The Dangerous Stranger*,[23] first released in 1949 and funded with help from John Wayne, directly cites Stroble before depicting children who made choices Linda didn't have, like taking rides with strangers promising ice cream and, in some cases, being saved because other children wrote down the stranger's license plate number.

Of course, Stroble was not a "dangerous stranger." His familiarity is what made Linda vulnerable. Knowing his license plate number was not going to save her. Just like it didn't matter much that Stroble was heterosexual, it didn't matter much that he was not a stranger to Linda. Davis's films were remade and shown to multiple generations of children. The cultural takeaway from this tragedy is an enduring trope: don't take candy from strangers.

The idea of dangerous strangers weren't invented by Sid Davis or introduced by Fred Stroble's violence, but these are among its earliest and most potent expressions. Fear of strangers is a major component in the weakening of community ties over the second half of the twentieth century. The shock of axe murder was secondary to the shock of learning to treat neighbors and strangers alike with suspicion. Parental (and especially maternal) expectations of surveillance have skyrocketed. But there is no way to eliminate risk from the simple joys of playing with your best friend; your neighbor probably isn't a predator, but there's no guarantee they won't grab the axe in the garage.

Felling Axe

1980, Wiley, Texas

By 1965, the axe's decline was undeniable. After two hundred years at the axis of American construction, labor, and domestic life, it was no longer an object of constant relevance. Where it was the once the first tool you'd reach for, now it was just one old tool among many. So many other blades on the wall of the garage, in the toolbox, in the barn.

Brett McLeod, forestry professor and author of *American Axe*, puts the shift down to a few interrelated factors. The chainsaw was the big one. Neither lumberjacks nor hobbyists were using axes to chop down trees anymore. Early-twentieth-century chainsaws made strides in the lumber industry, but they were huge, difficult to operate, and incapable of diagonal cuts in wood. In the 1960s, gas-powered chainsaws hit the market. These were much smaller, easier for transport and use by a single person, and much more appealing to the massacre-minded consumer.

And the way people were living was changing. Wood heating was on its way out, due to the pollution of its smoke and the labor involved in getting sufficient firewood. New houses in suburban tracts might be built with fireplaces, but that certainly wasn't the central method of heating the homes. The next generation was moving away from farm communities where they might be called upon to fell a tree and into the outskirts of metropolises, where the only use they had for an axe was to "grub [out] an ornamental shrub in the lawn."[1]

TWELVE

CANDY

Betty Gore was emerging from a rough patch when she was struck forty-one times with an axe.

In her youth, Betty Pomeroy was the prettiest girl in Norwich, Kansas. Her farmer parents thought she could do better than nerdy Allan Gore, her nebbish math TA at Southwestern University. She worshipped the quiet, intelligent man, so unlike the farm-boy jocks she'd dated in high school, and married him halfway through her sophomore year.

Marriage was a rocky transition for the couple. While Allan was pursuing his Ph.D., Betty had a one-night stand. She confessed to the infidelity immediately, and the couple stayed together.

After Allan finished his doctorate, the Gores moved to the northern Dallas suburb of Wylie so Allan could pursue the many opportunities on offer in the Silicon Prairie. It was basically what Betty had always wanted, but she wasn't thriving. Her pregnancies aggravated her anxiety and depression, and motherhood was not the balm she hoped it would be. She had a tendency toward moralism and aloofness with her peers, aided by deep insecurity and a natural tendency toward shyness. Thirty was approaching fast, and without the Kansas community that would pack the church at her funeral, she was lonely.

In Kansas, Betty had been popular, secure in the love of her parents and admiration of her peers. In Texas, she basically had only Allan, who was constantly on the road. And pregnancy had changed her body; she felt dowdy with a larger frame, and her insecurity inhibited her attempts at socializing. She found a teaching job but clashed with students and staff; she took in a foster son and found it to be a bad fit.

Betty was trying and struggling. She clung to her husband, hoping to force the intimacy that had never come easily to either of them. She was terrified of being left alone, and he was gone for weeks at a time. She turned to church in search of community and found her killer instead.

The Gores were brought to Lucas Methodist after their pastor, Jackie Ponder, alongside her close friend, Candy Montgomery, came to the Gore's Plano congregation to talk about the role of women in the church.

Jackie was a rulebreaker. "[Jackie] still plays cards and continues to be as feminine as before, goes to beauty shops, and wears eye shadow and nail polish," marveled Ellen Stone in *D Magazine*'s December 1976 issue. "She served champagne at her daughter's wedding and has been playfully admonished by parishioners for her use of certain unlikely words. But Jackie is not apologetic about any of it."

In her youth, Jackie had followed the same neatly paved path as any suburban wife: married by nineteen, occupied a nice house, had two kids. She came to regret skipping out on college and went back for a degree at twenty-eight. In her thirties, she had an intense spiritual breakdown, declaring herself an atheist and secluding herself from everyone but her husband. When she heard the call a few years later, it was a shock to everyone, including her new parishioners, many of whom left. Lucas Methodist was down to just twenty-five members when Candy first attended.

Candy had just moved to Lucas when she connected with Jackie. She had been married to Texas Instruments engineer Pat Montgomery for seven years in 1977, when they built a custom-designed dream home with a cathedral vibe that was perched on the crest of a hill in the countryside north of Wiley.

The term quarter-life crisis wasn't a thing back then, but Betty and Candy were going through it. Both were in their late twenties, younger than Jackie. They were wives and mothers with houses and kids and church responsibilities to look after. The excitement of youth and courtship were long behind them. Candy was a spitfire, used to demanding attention with her persistent cheerfulness and her petite-blonde good looks. She had grown up in France, and now she was just a wife in Texas, taken for granted by her husband and children. When Candy met Jackie, the freshness of her beautiful home was wearing off as quickly as new car smell.

Jackie moved into a small, dilapidated frame house behind the drafty church, at first with her husband, who soon after moved out. The two loquacious women bonded quickly, first through lively debate about the Bible and Candy's broadening role with the management of the church. Soon they were confidants. For the first time in years, Candy felt challenged—and appreciated. A voracious reader of bodice rippers and the pulpy historical fiction of Taylor Caldwell, Candy was suddenly discussing existentialism and German philosophers over coffee. She wondered if she too should get a college degree. Even though her family was more prosperous than ever, she felt trapped. Jackie's ability to make her own choices, whatever the consequences, looked a lot like freedom from Candy's perspective.

"I'm so scared about being single," Jackie said to Candy during one of their heart-to-hearts.[2]

"Jackie, you don't know how lucky you are," Candy replied. "And I am bored crazy."[3]

By 1978, Candy would talk frankly to Jackie about her dissatisfaction with her marriage. She was sick of sex with her husband; she wanted something exciting, like in a gothic romance. "I've done all the things a wife should do," she told Jackie. "The house, the kids, the fancy meals. And then one day I wake up and I say, 'where's the payback?'"[4]

Jackie wouldn't condone it, but wouldn't condemn it either. It was adultery, but she understood the pull toward new erotic experiences after years of increasingly tedious marital relations. She just warned her friend: "something bad will come of it."[5]

A former military brat, Candy had learned to charm people early in her childhood spent partially in France. So when she decided to embark on an affair, she did so with her trademark determination. When she set her sights on Allan Gore after they got to know each other through church volleyball, she knew how to get his attention. Candy was used to bringing out a "playful quality in men." She just slid right into his car and said, "I'm very attracted to you and I'm tired of thinking about it and so I wanted to tell you."[6]

Allan initially rebuffed her, but Candy was determined. They debated the merits of an affair via surreptitious phone calls and lunches: how it would affect their families (especially the newly pregnant Betty), whether it would be worth it, whether they'd be able to stop once they began. At one point, Candy even made a pros and cons chart to bolster her argument. In December 1978, they didn't so much as succumb to desire as take a calculated risk. Allan and Candy consummated their relationship in what would become a biweekly tradition: a home-cooked lunch at a motel, followed by sex.

Candy felt both satisfied that she'd pulled off an affair and a little underwhelmed. She wanted bodice ripping, and in her view the sex was conventional and gentle from the start.

The seediness of the Como Motel was exciting, though. The drooping curtains, the matted carpet, the red wine in plastic cups covered by Disney-branded cellophane. It was such a departure from the crucial work of keeping up appearances as a wife and mother. Even if Allan wasn't that attractive.

"He's not that handsome. He has a receding hairline. He's not what I would call my type. I don't know what I see in him," she admitted to a friend.[7]

After just a couple of months, Candy began to consider ending the affair. The novelty of a new body waned after a few encounters, and she suspected that Allan just wasn't capable of delivering the pyrotechnics.

Worse, she was starting to have feelings for him. She thoroughly enjoyed their conversations—about church, their spouses, their kids—and she showered him with affection. Even though the sex wasn't that great, she could talk to him in a way she couldn't with anyone else. Perhaps that's what she was really seeking out by choosing a not especially sexy man to have an affair with. She may have claimed to want an orgasm parade, but instead she looked for pretty much what she already had: an emotionally but not sexually satisfying relationship.

Allan kept up with Candy for a while after Betty gave birth. Until one night when Betty uncharacteristically initiated sex after Allan had had an afternoon tryst with Candy. Allan couldn't perform, which was a blow to his masculinity, and he broke things off with Candy. Candy felt it was unfair to end the affair. It wasn't a gothic romance, but it was comfortable. She loved the conversations they had during their fortnightly liaisons and relied on Allan's insight and humor.

And, throughout the affair, their families, once just church acquaintances, were becoming closer. Their daughters had become best friends.

Not long after the affair ended, the Gores went to a weekend-long seminar called Marriage Encounters—a more religious alternative to the fad of New Age personal development retreats like Esalen and EST. It was a major trend in the Lucas Methodist community; couples would go off for a weekend of presentations and journaling and alone time, then come back holding hands and spouting buzzwords.

For a while, Betty felt better. Allan's emotional resistance and the difficulties of pregnancy and parenting a toddler had caused her acute anxiety, especially with his heavy travel. "When I think of your being gone I remember those times with dread. The aloneness, the coldness of a house that really wasn't a home without you there, the fear for your safety, because you were where I was not, and I couldn't make sure you were okay," she wrote in a 1979 letter to him as part of a Marriage Encounters exercise.[8] The friendship she found through the group was healing for lonely Betty. Allan was finally done with Candy, and he was making an effort to improve the marriage. They planned a second honeymoon to Europe for late June 1980. Meanwhile, Candy had another affair with another man.

On Friday, June 13, 1980, Allan had to go to Saint Paul for a night. Betty was distraught about his leaving. She had a number of physical symptoms—shoulder pain, sinus problems, bladder infections—that flared up whenever she faced emotional distress, and she was especially distressed that day because of a late period and the prospect of an unwanted pregnancy. Betty tried to get Allan to skip his trip, stay with her, care for her. But Allan was determined to go.

He was able to cheer her a bit before he left. As he pulled the car out of the driveway, Betty raised their infant daughter's hand to wave goodbye and gave him a real smile.

Their older daughter had spent the previous night at the Montgomerys' for one of her frequent sleepovers, and she wanted to stay another night. But she had a swim lesson, and so Candy needed to pick up a swimsuit from Betty.

A little before nine, Candy took the kids to vacation Bible school at Lucas Methodist. Before their nine-thirty lesson, she told a story about a woodcutter to the kids. Chopped limbs fresh in her mind, she set off to retrieve the swimsuit from the Gore house in Wylie.

This much we know from various secondary sources. But the last minutes of Betty's life have only one narrator. Candy.

Betty wasn't expecting her until noon, so she was surprised and a little irritated by the early arrival. Candy put on her friendliest smile, and asked if they could extend their daughters' sleepover by an extra night.

Betty showed Candy into her messy living room. Phil Donahue was on. Betty was dressed in house clothes: old and ill-fitting shorts and a yellow blouse. She didn't expect to leave the house or host people that day. Betty turned off the television and sat down. As usual, she didn't have a lot of energy. But she made an effort: they chatted about the baby, swim lessons, the Gores' new puppy, their trip, Candy's new home-decor business.

Candy implies that Betty was perhaps a little jealous as Candy showed off her new business cards. The Cover Girls, Candy and her friend had called it. Cute, as always.

Up to this point, both of the women seem to be following their usual routines: Candy, friendly to a fault; Betty, nervous and a bit overwhelmed. Through the years they'd known each other, Candy had always been a step ahead, and not just in carrying on an affair under Betty's nose. Betty wasn't one to outwit her friend or assert herself.

But at this point in Candy's story, Betty makes a startlingly accurate connection with no apparent provocation, months and months after the affair's conclusion. Even though her brain was in a fog from the stress of an infant and an absent husband, even though she wasn't prone to dead-on social calculations, Candy wants us to think that Betty just suddenly figured it out. "Candy, are you having an affair with Allan?" Betty asked, as boldly as Candy had once asked her husband to initiate that very affair.

"No, of course not."

"But you did, didn't you?" Betty's accusation had an unfamiliar aggression.

In Candy's telling, Betty was no longer tired and fuzzy, but steely, eerily calm. "Yes, but it was a long time ago," Candy confessed, cowed. "Did Allan tell you?"

"Wait a minute," Betty said, and went to the utility room. Candy sat at the table, waiting.

Betty was a timid woman, everyone agrees. She could get dramatically emotional, and she was sometimes overzealous in the enforcement of rules, laws, vows. But Betty was one to always follow the established path. Betty was so upset by Allan's absences because they made her life feel unmoored, unpredictable. She tried to schedule her pregnancies to the week; she had probably planned the European vacation that she was so looking forward to down to the very minute. She wasn't one for spontaneous acts of anything, let alone violence.

In Candy's narrative, Betty reacted to freshly devastating

news not with blubbering tears but by procuring a weapon of convenience from time immemorial: the axe.

"I don't want you to ever see him again," she said, like a character from a cheesy and obvious slasher movie. Candy said Betty held the axe awkwardly; it was a tool used for chopping wood, with a three-foot handle curved to allow for a better grip and an angle to split logs. "You can't have him."

"Betty, it's been over for a long time. I'm not seeing him. I don't want him," Candy said. She waved her hand, like she hadn't begged Allan not to dump her. "Don't be ridiculous. It was over a long time ago."

As Candy tells it, this defused the situation for a moment. Enough so that she actually stayed in the house despite being threatened with an axe. Even in her own story, she had multiple opportunities to just get the hell out of the house. I don't think it's too big a stretch to say that any reasonable person who didn't mean any harm would have left at this point. Not Candy.

Candy's story is that she didn't initiate or escalate the violence; at every turn, Betty escalates, pushes, swings first. But for me, the story collapses when Candy stays in the house despite repeated opportunities to leave and get away from a woman she paints as clearly unstable and dangerous.

Instead, Candy says that after Betty's armed threat, they returned to the quotidian details of the sleepover. Candy said that she'd bring her daughter by right after vacation Bible school; Betty said no, that she didn't want to see Candy again, and that the Montgomerys should keep her for the night and bring her back tomorrow.

Betty put the axe against the wall and went to get a towel for Lisa's swimming lesson, while Candy waltzed past the axe to retrieve Lisa's swimsuit from the utility room. Betty fetched some peppermints for Alisa while Candy waited, fidgeting with her purse. Betty fell into tears in the hall.

"Oh Betty, I'm so sorry," Candy cooed.[9]

And that is when this Betty, full of that uncharacteristic aggression, grabbed the axe and shoved Candy back into the utility room, yelling that Candy couldn't have Allan.

Candy peppers dialogue throughout her story. Betty, robotic and terrifying with a distance to her voice, Candy begging for her cute little life. Straight out of a pulpy novel, which is not an insult to pulpy novels:

"Please don't!"[10]

"I'm going to kill you."[11]

"Let me go, I don't want him."[12]

They struggled for the weapon, eyes locked. Betty jerked the axe toward her, and it rebounded to hit Candy in the head. Betty raised the blade above her head and brought it down; it missed its mark, but bounced off the linoleum and hit Candy in the toe.

The women wrestled on the floor, Candy said. The dogs howled outside. Betty bit Candy's knuckle in an attempt to get leverage. Finally, Candy got the axe and hit Betty in the back of the head. But she survived the blow and, like a zombie, threw herself against the utility room door before Candy could get through.

Candy tried to get to the door to the garage. Blood was pouring out of Betty's head, but she managed to get there first, again. Candy recalled that the woman with a brain injury even had the presence of mind to lock the door.

In Candy's memory, Betty raised a finger to her mouth. "Shhhhhh," she intoned, eerily of course.[13]

And then Candy destroyed Betty's face.

Candy didn't return to the church until noon. A church friend had expected her back at eleven; Candy told the friend that she and Betty got to talking, and she lost track of time because her watch stopped. That evening, Candy and Pat and their children and Betty's daughter all went to see *The Empire Strikes Back*. Afterward, they had tacos and the kids played in the yard together.

A neighbor checked in on Betty after Allan couldn't reach her, and found Betty's infant daughter crying in a soiled diaper, as she had been for hours. Betty's body was in the utility room. Her head was so destroyed that the neighbor told Allan that Betty had shot herself. But she hadn't. Concealed beneath the fridge was a three-foot axe covered with her blood

Police began their search for an axe murderer. They were looking for a rampaging maniac who would kill a woman with an axe in her home in the middle of the day with no apparent provocation. Maybe it was a cult deal, inspired by the homicidal performances playing in cinemas (there were two of them: *Friday the 13th* and *The Shining*). Maybe it was a spurned lover with a real motive, or a sex crime—it was clearly a crime of passion, the way Betty's face was mangled.

Medical examiners figured it had to be a man, given the

force necessary to break Betty's cranium and dig into her brains. The damage to her body was so extensive that it was difficult to look at her directly. Someone must have really hated her to have struck her so many times while her heart was still beating, they thought. And in 1980 the axe wasn't something a person used for a premeditated crime except in movies.

"People don't use that very often to murder someone," the investigator on the scene said decades later.[14] Guns won the battle long ago.

Candy was the stuff of no one's nightmares, and this case was more than a nightmare. These were the suburbs. White people don't hack each other to death in their secure, well-maintained homes on a Friday morning. It was as verboten as letting your grass grow a foot high.

Everyone knew that Candy had been the last to see Betty alive, but no one considered the petite woman a suspect in a murder this vicious, committed with an enormous axe like that. People believed Candy when she said that she just picked up the swimsuit and then headed for Target to get a Father's Day card for Pat. Her wounds were easy to conceal: a cut at her forehead, a gash on her big toe, some bruises. She combed her hair a little different, put on tennis shoes instead of her trademark flip-flops, and told Pat that she caught her bloodied foot on a screen door.

The police thought she was a big help with reconstructing Betty's last morning. But otherwise, they were stuck. The prime suspect, Allan, could not have had a better alibi. He was literally a plane ride away, surrounded by colleagues the entire time.

He was guilty of being a poor husband, though. While police searched for Betty's killer, Allan was efficiently going about the business of forgetting all about her. At his wife's funeral, he realized he felt no real grief for her. He even told Betty's father, Bill Pomeroy, that his wife's violent end "didn't bother me very much. We weren't getting along anyway."[15] In the weeks to come, he would take up with a neighbor lady who subjected the girls to terrible physical and emotional abuse, going so far as to make them give book reports about their mother's murder; eventually, Betty's family got custody of them, thank God.

Allan did have a conscience; it wasn't very reliable, but it was there. Five days after the crime, he called the cops and told them what he'd left out of his story: his affair with Candy Montgomery. Of course, he was quick to defend her. She loved Betty! Candy didn't wish her any harm, according to her former lover. But the police chief wasn't quite as trusting, especially after Candy had failed to mention the affair.

After one interview, Candy was smart enough to know it was time to get a lawyer. She hired Don Crowder, a tough-talking attorney she knew from church and had occasionally flirted with. During her initial consultation, she didn't say anything and he didn't suspect anything. She was cracking jokes.

Besides, Candy was a cute little housewife. There was no way she'd find the edge against a much bigger woman.

At first, Don didn't believe it when Candy told him the truth: she did it. It took her a full seven days to actually say what had happened that day; she hadn't even told Pat. Violence can do funny things to a memory, and it usually doesn't make it more reliable. And Candy was a natural charmer, a decorator who always knew how to put things in the correct light. But Don came to believe Candy's story, and believed he could win her freedom with it by claiming the act was self-defense and mental defect in forty-one blows.

As he told it at court, Betty started it, and Betty was fat, and sweet little Candy was terrified by her fatness. So she had to hit her with an axe forty-one times, ripping off her face and crushing her skull. Allan was no help, as per usual; he couldn't be bothered to gesture at grief for the jury. Even he seemed sorry for Candy. Poor Candy was already in a cell—losing her house, savings, respect of the community. Candy had earned her status symbols, damnit. Sure, dying violently is awful, but the trauma experienced by having killed someone was surely punishment enough. Crowder brought out the axe Candy had used to kill Betty and pushed it at her. She cried and cried and cried and cried for the jury, and they seemed affected by her display of emotion.

But they had to explain the forty-one whacks she dealt to Betty. Overkill needs a really good rationale. A psychiatrist who performed hypnotism on Candy provided one. Candy, the formerly unrepentant adulterer, had been brought to the brink of insanity by her own desire to be good.

You see, in the course of their confrontation on the morning of the murder, Betty had shushed Candy.

This, the psychiatrist testified, made Candy regress to an event she experienced at age four, when she lost a race to a neighborhood boy and smashed a glass in a fit of pique; the glass cut her, and she had to go to the emergency room for stitches. She was very upset, kicking and screaming, and her mother shushed her, asking "what will the people in the waiting room think of you?"[16] This thought was so excruciating to Candy that it led to over twenty-plus years of repressed anger, which exploded in brutal violence. The fact that it exploded at Betty didn't have much to do with the fact that she was a romantic rival; Candy had simply been under too much pressure to be the perfect Texas wife.

Improbable, but it works. At that moment in American history, especially; juries across the country let all kinds of wild Twinkie defenses fly in the late seventies and early eighties, which would change after John Hinckley Jr.'s acquittal in 1982. The timing was right for Candy to get away with murder by mounting a defense that gestured at trauma.

The psychological harm done to a shushed four-year-old is small measured against the scale of the damage to Betty's body. But I'll give Candy some benefit of the doubt, as any accused party deserves. She didn't go over there to kill Betty that Friday. I believe that Betty was the party who introduced the axe into the conversation, and likely the one who turned the confrontation physical. And yet—forty-one whacks.

But the proportions of the crime made sense to the jury, who returned a not-guilty verdict in short order. In the eyes of the law, Candy was no axe murderer.

Candy and Pat moved the family to Georgia, and she went on to become a therapist.

* * *

Even more than the axe, it's the absolute gall of Candy that made this story ripe for its three television adaptations. The shocking use of an obscure childhood memory to justify Candy's violence is anticathartic: there is no relief of justice, just the reminder that people get away with atrocious nonsense all the time. Perhaps the lack of relief is why, in many of these adaptations, Betty comes off so poorly—cloying, desperate, downright mean.

I felt as if the killing had been mostly forgotten when I first started writing about it in 2019. There was a 1990 television movie, *A Killing in a Small Town,* that focused on the courtroom drama more than the domestic violence; I watched it on YouTube. Candy didn't even have a Wikipedia entry at that point. But the recent true-crime boom, driven by women, came for the story right around when I did, with two adaptations on streaming services a year apart.

The 2022 Hulu adaptation *Candy* is seedier, lowbrow, a little more "TV movie." It features the axe much more heavily as a motif; the 2023 HBO adaptation *Love and Death* barely touches it until the actual crime. But in its higher-mindedness, the HBO adaptation somehow manages to fall for Candy's story again. *Love and Death* luxuriates in the period details, resembling a 1979 version of *Big Little Lies* (produced, like *Love and Death,* by David E. Kelley). The camera adores Elizabeth Olsen as Candy, who is constantly taking showers. There are multiple scenes where she gratuitously overreacts to shushing, the hypnosis story is presented as basically reliable, and her version of the events of the murder is presented without interruption. The theme features Nina Simone singing "dear Lord, please don't let me be misunderstood" as Candy chops absently; the show takes her too seriously, wants the viewer to acquit her all over again.

The Hulu version, with Jessica Biel as Candy and Melanie Lynskey as Betty, is wise to Candy's cunning. It's an ashtray-gross version of the late seventies, flat and more than a little ugly. Perhaps it's a little heavy on the dark comedy, but it also sees the absurdity of Candy's explanation. Perhaps it's too hard on Betty, but when the death is finally depicted in the final episode, she is there as a scowling ghost in the audience as Candy performed in court. Perhaps it's a little too obvious, as when Pat counts out forty-one whacks in the backyard during the trial, but it also points out the obviousness.

It's all based on the reporting of John Bloom and Jim Atkinson, first in *Texas Monthly* and later in a quite good book titled *Evidence of Love*. The book is based on extensive interviews with the major figures in the story, especially Allan and Candy. Though they also rely on Betty's childhood diaries and letters to Allan, she wasn't around to give her perspective. And it shows. *Evidence of Love* doesn't exactly endorse Candy's and Allan's behavior, but it renders their thought processes, their motives, their explanations. It's not so much the story of Betty's tragic death as the story of how Candy came to kill her—and how she avoided conviction.

It's a problem that Bloom and Atkinson can't fix, and neither can I. Betty died so long ago that her daughters barely remember her. The narrators of her death were her husband and her killer, both selfish people with a strong motive to make excuses for themselves and their actions, which are considerably more despicable than—what? Asking for more from your husband? Seeking friendship and community at church?

Betty wasn't perfect. There are some verifiably negative things that she did. She was quite harsh as a teacher, and she was not up to the trials of fostering a young boy. She had a desire to nurture but perhaps less of the actual instinct. She probably was a bit awkward, a little needy. But I just don't think she was always like this; when she married Allan, she was popular and desired. There was a journey from being the prettiest girl in a farm town in Kansas to being the desecrated body in the utility room of her house in suburban Dallas. And yet the story is never about that journey, not even here. Because her killer and her unfaithful husband live to tell the tale. The story is always going to find its way to a narrative that blames Betty because Allan and Candy couldn't face what they did squarely. They've never had to do so, and television adaptations aren't going to solve that problem.

"I just think she got away with it," Betty's younger daughter said in an interview with the *Dallas Morning News* in 2000. "I wonder if [Candy] thinks about it every day, like I do. I wonder if she thinks about us."[17] Betty's daughters had a difficult period after Candy's acquittal; Allan and the neighbor he quickly remarried were abusive. Eventually, the girls moved to Kansas with Betty's parents, where they had a relatively decent childhood. But of course they did not come away unscathed. "I just wish I knew what really happened, because nobody knows but [Candy]."[18]

Fire Axe

1980, The Overlook Hotel

On Betty Gore's kitchen table that Friday the thirteenth, she left her paper open to a review of *The Shining*. Stanley Kubrick's film adaptation of Stephen King's 1977 novel opened on May 23, 1980. *The Shining* wasn't a hit yet, though it would be by the end of the summer.

When my husband, Jason, saw the movie with me for the first time early on in this project, he was struck by how many pop culture references to the film he now newly understood and appreciated—the empty bar animated by a delirious dry drunk, REDRUM, all work and no play, the final close-up of the sepia photo, the Grady twins. There's barely a composition, a transition, a set-design choice that has not become iconic in the forty-five years since its release. There are four different iconic bathrooms in that movie. Have you ever noticed? When we got to the most famous one, in which Jack Torrance hacks his way through the bathroom door intent on murdering his wife, Jason started laughing.

I don't think he intended to confirm my hypothesis that axe murder has gotten funny. But he has seen this very scene so many times before he actually saw it, almost always in parody. This one scene is a major factor in the continuing legacy of axe murder in comedy. Though he tends more toward Nintendo than *Grand Theft Auto*, he's as desensitized to images of violence as any elder millennial.

Desensitization is not the only element at play here. There is something grimly zany about the climax, Jack Nicholson's wide grin a preposterous contrast to his awful reality, the axe blunt evidence of the Overlook's terrifying hold on his imagination. It's the humor that makes it so shocking—and memorable.

Stephen King famously hates Kubrick's film. He protests that Wendy is a character so weak she borders on misogynistic and believes that the film is not sympathetic enough to the abusive drunk, Jack Torrance. Perhaps his biggest beef is with the ending: "The basic difference that tells you all you need to know is the ending. Near the end of the novel, Jack Torrance tells his son that he loves him, and then he blows up with the hotel. It's a very passionate climax. In Kubrick's movie, he freezes to death."[1]

King has some points. Kubrick is far less interested in the family backstory. The grandly terrifying visuals take priority over both the cyclical trauma suggested by Wendy's and Jack's own shitty upbringing and insight into Jack's mind as he's drawn further into madness. But regardless of what King himself thinks, Jack's final act is no redemption. It doesn't really matter that Jack told Danny he loved him at the end of the book. Abusers claim to love their children all the time, even while they are actively abusing them. Kubrick, with all his coldness and black comedy, understands Jack Torrance on a simple and essential level: this is a bad man.

The axe he carries in the ending is not just a visual upgrade from the mallet he carries in the book. It's a visual symbol of Jack's active participation in the horrors of the Overlook, a choice he makes for himself.

The axe he takes, or with which the hotel supplies him, is a firefighting tool. The spike on the end is for punching through vents and steel siding. Fire axes were installed in hotels because of the inherent danger fires present for hotels, even those unpossessed by ghosts. The life he chooses to save is the Overlook's.

The review Betty read that Friday in June 1980 probably wasn't positive. Film critics from Roger Ebert to Janet Maslin to Pauline Kael hated the film when it opened. *The Shining* was a commercial success, but it took most of the decade to earn a critical reevaluation, all while Kubrick's film became more and more essential to pop culture. It was a far less successful film, however, that made the axe murder an overt punchline: *So I Married an Axe Murderer.*

Robbie Fox's script for *So I Married an Axe Murderer* was originally sold as *The Man Who Cried Wife*. That's when the perfectly named Dawn Steel got ahold of it. Steel was the head of Columbia, the first woman ever to run a major studio. In a meeting with Robbie, she told him that she didn't know what his script was exactly—slasher? romantic comedy? thriller?—but she had an instinct about how to make it a success.

Steel "threw it down on the table," Robbie recounted to me, and she said, "Let's put the axe murderer in the title, and let's make it.'"

Robbie was pleased with the title change, but not the other adjustments yet to come. When Mike Myers came on board, it changed the tone that Robbie had envisioned. The basic plot was intact: a commitment-phobe meets a free spirit who seems like the one, until a tabloid true-crime story makes him fear that his new wife is a serial husband killer plotting to murder him (spoiler alert: she's not, but her sister is). Robbie wanted a thriller with a comic edge—"essentially Woody Allen in Hitchcock's *Suspicion*."

"There was never an axe murder until [I was] scared I would get fired," he told me. Then, divine inspiration struck in the form of a postcard that producer Robert Fried got from John Landis, vacationing at the Banff Springs Hotel, a grand hotel in the snow-coated mountains of Canada. On the back was written, "I feel like I'm in *The Shining*!" Fried told Robbie he had the ending for the movie.

"I went and watched *The Shining* and decided to replicate

that," Robbie said. "I'm just gonna literally re-create Jack Nicholson chasing his family down the hallway with an axe.

"It's just a clunky, heavy weapon. An axe is so unfeminine," he added.

Once *Axe Murderer* was a Mike Myers vehicle, it acquired his particular style of knowing, deconstructive, gag-heavy comedy. Plenty of Scottish accents and digressive satire and outlandish set pieces—not a lot of subtlety.

Audiences at the time responded poorly: *So I Married an Axe Murderer* was a huge flop, netting $11.5 million at the box office against a $20 million budget. Yet the movie has had a long afterlife and is now revered as a cult classic. "It inverts the priorities of your average '90s rom-com, placing camp and dark humor at the forefront without sacrificing the endearing romance at its center," wrote Maggie Serota in *SPIN*, in a retrospective appreciation.[2]

After *Axe Murderer*, the phrase became a common punchline among similarly inoffensive movies and shows of the late-twentieth and early-twenty-first centuries. A joke that was a bit edgy in 1993 has since become so anodyne as to be imitated for corporate events like "So I Hired an Axe Murderer: Learn to Minimize Risks in the Hiring Process," a forty-dollar session hosted by Suncoast Healthcare Executives in January 2020.

The phrase is all the more anodyne because it is meant increasingly as an anachronism. Rightfully so. Where the axe was once essential, its uses have narrowed so much that there's usually another, more efficient tool available, whether for woodcutting or murder.

EPILOGUE

Crime and Difficult Situations

Kansas City, Missouri, 2019

When Mario Markworth was four, his five-year-old sister called 911 to ask for a pizza because they hadn't eaten in two days. This was a case where state assistance was absolutely necessary, but the Department of Family Services in Kansas City, Missouri, only made things worse for Mario. The siblings were separated, and Mario went to several different foster homes in which he was exposed to "crime and difficult situations."[3]

From 2000 to 2004, Mario was neglected, starved, and sexually abused in the care of the state. He was even deprived of water and had to drink out of the toilet bowl.

In 2004, Mario and his three siblings were adopted by Doug and Terssa Markworth, a couple in Nebraska who count themselves as parents to twenty-one children, sixteen through adoption. His life circumstances improved immeasurably from that point. Terssa cared deeply about Mario as a person, and wanted to give him as good and loving a childhood as they could; he was safe, fed, loved, protected from harm. He had merry Christmases and happy birthdays.

The Markworths were Mario's salvation, and they enjoyed that feeling of goodness more than anything. Doug and Terssa met in the air force when they were barely adults and married after three months of dating. From the beginning, Terssa said she wanted ten children; they started out with four biological kids, all sons. After Doug got a foothold in his career with State Farm, they tried foster care. They found it too painful to return the children to their families, and so they began adopting groups of siblings, sixteen altogether. They did not consider race in their prospective adoptions, for good and for ill; Mario and his biological siblings are black. Terssa's 2015 self-published memoir *You Can't Bungee Jump with a Pair of Pantyhose!* does not use the word "racism" once, and at one point she scoffs at the idea that her own white privilege could negatively affect her children who are Asian, Hispanic, or, like Mario, black. He was on the honor roll but felt isolated and alienated in high school, and after graduation he headed back to Kansas City to be near his sister Brittney.

Brittney blames herself for what Mario did alone one August night in 2019. He had lived with her for a few years by then, acting as a loving and protective uncle to her children. She suggested that it was time for him to get his own place. It was the kind of sisterly wisdom any parent would hope their children might share with each other. She thought he was ready for the challenge, that he wouldn't get into any real trouble; she never saw him fight with anyone. Twenty-three is a perfect age to start taking more responsibility for yourself.

He moved into the Greenleaf complex, an unremarkable set of eleven buildings, with brick in front and faded white vinyl siding in back. There's a little playground in the middle. Right next door is a squat little place with glass brick windows, lime-green doors, and a light-blue and black diamond motif on its trim called Scandalo's Party Hall.

Mario found a community at the Greenleaf, but it wasn't the kind Brittney hoped for. He felt "picked on" by the other residents there and pressured to participate in the party scene.[4] Other than marijuana, Mario didn't get into drugs while he was living with Brittney, but his substance use increased dramatically once he moved out on his own. It was in this situation that he met Michael McLin, an older man who was also in the Greenleaf party scene. The two Kansas City natives knew each other well enough; Mario called him Uncle Mike.

At the Greenleaf, Mario was no longer feeling stable or safe. His sobriety, his funds, his living situation—they were all a little tenuous. To compensate for this feeling, he got an axe.

We know from the police report that witnesses saw Mario carrying the axe around on the day of Monday, August 5, 2019. Perhaps he purchased it at a hardware store; axes are not expensive. Perhaps he found one somewhere and thought he'd keep it just in case.

I don't think he meant to use it. There was no planning

beforehand. If Mario wanted to kill two people, he would have set about getting a gun. Firearms aren't much more difficult than axes to obtain.

Mario and Mike left the party to go get alcohol and other supplies at a nearby convenience store. On the way, they passed the Sheffield Family Life Center, a Pentecostal megachurch with a huge parking lot of seven hundred spots. A megachurch parking lot is as close as you can get to the middle of nowhere when you're actually at the center of a major city. And a lot of secretive sex happens in parking lots.

In the police report, Mario stated that they agreed to "play with each other." They were alone after a night of drinking. It's probably after midnight and still above eighty degrees with humidity at 70 percent. I don't know if it was their first time hooking up. There was a big age gap between the two men, but they were both adults. Mario was naked and Mike's pants were off when Uncle Mike changed his mind.

Mario freaked out over Mike's rejection; he was afraid Mike would tell someone about it. He was terrified of how wanting to have sex with Mike would change his reputation at the Greenleaf. The imagined mockery from peers who bullied him, the disappointed expectations of the family members who loved him. When Mike started to walk away, Mario picked him up and threw him to the ground, where he said that Mike's head hit the axe. Mario picked up the axe and hit Mike in the head four times to finish the job.

Michael McLin's head had seven incised wounds, according to the testimony of deputy medical examiner Lindsay Holland. They were longer than they were deep, usually the kind you'd see from knives. It also had four "chop wounds," or "sharp-force injury."[5] Two of the chop wounds were especially brutal: one went all the way through the skull, and the other through the left carotid artery to the backbone. There were fifteen injuries to the torso as well. Mike was found face down, covered in blood and debris, his pants around his ankles.

Mario's motive—the rejection, the shame—was in the police report, but it was not mentioned at the court hearing. No one brought it up, for the same reason that Mario felt compelled to hide it in the first place. The threat of homophobia was deadly enough to incite violence. And the bloodshed did not end with Mike.

The last time Kevin Waters saw his son, Joshua was concerned about his father. He'd been sleeping outside for months; this wasn't a new thing for Kevin, but the August heat and the inherent vulnerability of having no door to lock worried Joshua.

Joshua "told [his father] 'don't let them kill you out there.' Waters told [Joshua] he would be fine."[6]

Like Mario and Mike, Kevin had a difficult journey through life, spending time in foster care in childhood and struggling to find stable housing in his later years. Through the trials of his fifty-two years, he kept a strong faith and maintained loving connections to his family and friends; they remembered him as a generous and loving man who lived a fearless life on his own terms.

Kevin grew up in Kansas City. He attended high school but never graduated; on Facebook, he listed his education as "School of Hard Knocks, The University of Life." His home-life was rough, and he often ran away from wherever he was staying. Throughout his life, he consistently chose the freedom and challenge of sleeping rough over the convenience and ease of a home with someone else's rules.

"He called himself a hobo," Kevin's brother Tim Waters told Glenn Rice of *The Kansas City Star* in the wake of Kevin's death. "He would say, 'Tim, I know I have a family. I know I have a home but I want to be free.'"[7]

But even though he valued his independence, he didn't shirk his responsibilities. "He took care of his kids like no other," his daughter, Elizabeth, said.[8]

"He was one of the best human beings I've ever met, and he was homeless," Joshua said. "Being homeless doesn't diminish your value. It doesn't make you less than anyone else."[9]

In his last year, Kevin's Facebook profile showed a man who was struggling but kept a positive attitude. He found great joy in the Kansas City Chiefs; 2018 was Patrick Mahomes's first year, and they went to the playoffs for the first time in a long while. But he was at a low point when his tent and personal possessions were destroyed. He did have someone to lean on—his ex-fiancée, a woman named Denise whom he met at the Salvation Army. But they broke up, not for the first time, because she was on her meds and he was not. Still, they loved each other, and she helped him out when she could; they were in daily contact.

For a time that winter he had a place at the Samantha Heights apartments in Independence, but by spring he was sleeping outside again. He asked for prayers to get back on his medication and for help getting a new pair of glasses. In July, he wrote, "Thank you Jesus for the sun going down man it was a hot one today God bless be safe."

"I lay here looking at the stars and one goes out. Hmm, I wonder, how do they change the light bulb?" he wrote on his fifty-second birthday on August 1. "Good night Peeps. God bless and Be safe."

According to his Facebook profile—a place where he did not hesitate to share his problems—Kevin was in a pretty good place the week before he died. He posted a lunch (always the best use of social media) on August 3—King's Hawaiian rolls, a salad, an apple, Tic Tacs, and a Mountain Dew. On August 4, about thirty-six hours before his death, he posted several photos of lovely green views, a tent pad, a stone wall, stairs on a wooded trail captioned "Path of our lives."

In the early morning of August 6, he slept at the edge of the empty Sheffield parking lot. He was at the fence line, which "was lined with bushy trees and was largely concealed by overgrowth from passersby." There was something of a path along the fence line, which the police report noted was "compacted from regular use and traverse."[10]

Mario was moving Mike's body and hiding his white shoes when he noticed Kevin. Kevin was sleeping. Kevin had nothing to do with it. Kevin wouldn't tell the people back at the Greenleaf about anything. But Mario was not ready to stop.

Kevin got to his feet and fought for his life. They grappled for control of the axe in the bushes behind the church. Mario bit Kevin on the back and got the axe, and hit him with it many times. Kevin's body was more extensively disfigured than Michael's because their fight was more evenly matched and more intense. He had bite wounds on his back and seven defensive torso wounds. His head bore a dozen incised wounds, and five chop wounds, all of which fractured the skull. Three trespassed into his brain.

The next morning, a tow-truck driver found two bodies in the Sheffield Family Life Center parking lot. Later that day, a witness reported that Mario had been walking around with an axe "similar to what firemen carry" and fairly openly confessing to the crime. He was arrested the next day and admitted guilt immediately.[11]

In his possession was the axe. It was more of a hatchet, painted black, with a blunted butt. It was about eighteen inches, needing only one hand to swing.

Mario technically had housing, but his situation wasn't stable. In the face of homelessness, an axe is a practical tool. So much of the move away from axes is about improved housing circumstances: no more wood fires, no more butchering, no more clearing the land. Unhoused people need to do all those things, and they don't need to carry around a chainsaw. In a situation where having resources at hand matters more than just about anything, the handiness of the axe is once again powerful.

Then there's the other cheek. Surviving outdoors can lead to a lot of scary and desperate situations. Resources are scarce and conflict is intense. When the axe is at hand for cutting wood again, so is it at hand in a fight. The axe is the least relevant it's ever been, and yet it is still so common, so handy, so short.

Many of our violent protagonists chose the axe with forethought and care and even symbolic intent. More of them grabbed the axe because it was there to be grabbed and it was heavy and sharp enough to do the damage they wanted. It wasn't something he found on the asphalt in the heat of the conflict, left behind by church maintenance staff or another person sleeping rough. Mario kept the axe as a way to feel safe and powerful in an uncertain and sometimes hostile place.

I believed Mario when he said in court that there are many ways that tragic awful night could have played out differently. I don't think that he imagined using his axe on Mike when they set off for the convenience store. It's possible that the axe fell out of his backpack when Mario disrobed for their aborted hookup, and if it hadn't fallen, the fight wouldn't have escalated past a scuffle. The axe was just there. Nothing could have been simpler.

"Markworth stated he was raised in a Christian home and knows right from wrong and knows his actions were not right."[12]

So ends the police report concerning the arrest of twenty-three-year-old Mario Markworth on August 7, 2019, the day after he murdered two men with an axe in a church parking lot. Mario confessed to detectives with the Kansas City, Missouri, police department as soon as he was arrested with the weapon in his backpack; he never maintained his innocence or claimed self-defense or expressed anything but remorse for the terrible things he did.

In his last plea to the court, Mario's lawyer emphasized that they were not asking for a slap on the wrist, that there was no childhood abuse that could provide an excuse for his actions. He said he knew he wouldn't get the minimum, but he asked for mercy.

On the day he was sentenced to twenty-eight years in prison, he wanted people to understand how he came to be an axe murderer, which is why he explained the difficulty of his childhood. Through his sobbing lawyer, he gave a long statement in which he repeatedly emphasized that he did not want to evade responsibility for his "unforgivable actions." But the only thing he said when he turned his tear-stained face to the people who mourned Michael McLin and Kevin Waters in the courtroom on June 24, 2022, was "I'm sorry."[13]

There were about a half dozen of Michael's family members in the courtroom the day Mario was sentenced for his murder. One man brought Michael's adorable grandson, who gurgled and smiled at everyone in court. Michael's aunt Ethel, a trim woman in a tracksuit and floral top, testified that they would fall out from time to time, but that she never thought she wouldn't get the chance to say goodbye. His cousin Shelley, a pretty, apple-cheeked young woman with twists in her hair, said that he was a good man, especially before drugs got ahold of him; Michael was someone who "had his issues" but was ultimately "harmless."[14] She spoke of her sympathy for Mario's family, broken too, and expressed forgiveness toward Mario himself.

"Nobody wins," she said. "What's done is done."[15]

Kevin's family too said that they had no hate for Mario; on Facebook, I was struck by the way Kevin's stepsister referred to Mario only as a "gentleman." But they were still upset, not ready to forgive. The intensity of Mario's violence made these men's last moments a brutal terror. It is not the way anyone should die.

"Mario will think about [the murder], but he'll eat, talk, walk, enjoy birthdays and holidays," said the prosecutor in his closing arguments. "Kevin will not."[16]

Perhaps the most wrenching moment of Mario's sentencing was not when he said that he was sorry, but when Brittney did the same. She told Kevin's and Michael's families that she was sorry not on Mario's behalf, but on her own. The axe murderer's sister, who wasn't there that night, who tried so hard to help him, sobbed as she talked about the responsibility she felt for Michael's and Kevin's deaths and for letting Mario's life get to that point. She wept in the hall after she stepped down from the stand.

After the judge handed down his sentence, he dismissed the rest of the courtroom so that Mario could have a moment with his family before he confronted the next twenty-something years of his life. As Michael's aunt Ethel left the courtroom, she swept Brittney into a hug. They experienced different tragedies, but in their grief they were for a moment united through the generous refuge of forgiveness.

In *Crime and Punishment*, senseless violence is followed by a moment of meditation. Immediately after Raskolnikov murders two women, he loses himself in cleaning and examining the axe he used "long and attentively." The act is one of purification, not unlike Nordic rituals bringing down an axe around a wound without piercing the skin to cure infection. The catharsis of the act brings Dostoyevsky's antihero toward the realization of his madness, that "he was perhaps not doing at all what he should have been doing."[17]

The destruction and the punishment and the forgiveness we offer each other are often out of balance. For millions of years, our axes have been put to causes of justice and devastation and sacrifice, the tool a symbol of war and peace and home. The axe murderers in these pages are enthralled by machinations that took decades to manifest: the urgency of war, the dehumanization of slavery, the entitlement of royalty, the chaos of lunacy.

The axe holds none of these qualities inherent in its flint, bronze, steel. Its powers of construction and destruction are imbued not by the leverage of the handle or the sharp edge of the blade but by the one who uses it. Bringing it down upon wood or flesh is at the sole discretion of the lumberjack or murderer. Axes hung on the wall of a palace, gathering dust in the basement, or abandoned in the forest cannot alone draw blood.

ACKNOWLEDGMENTS

I want to start by thanking my husband and mailman Jason Graham. He was a constant source of support and understanding throughout many all-nighters, and made sure to show me every axe he equipped in every video game he played. He also spent a lot of time helping me get to major research sites, though we did fit in several Phish shows around it. The process has turned him into a bit of a Juggalo, but to that I say whoop whoop. I love you!

My agent, Laura Usselman, has understood this weird niche project that wasn't true crime but wasn't traditional history since we first spoke about it in 2018. This book absolutely would not have happened without Laura. She's offered a wealth of detailed editorial insight and business know-how throughout the process. I also want to mention her colleague Ross Harris, who came up with the brilliant title *Whack Job*. The title instantly created the exact energy that this book demanded—I can't tell you how many people have said "wow, great title!" when I've bragged about it over the years.

My editor, Hannah O'Grady, got what I was going for from the beginning, encouraging my sharpest insights while gently warding me off my worst (and most online, lol) impulses. Madeline Alsup dealt with so many book tasks with such kindness even as I was frustrating myself. Copy editor Angela Gibson saved me some headaches with her eagle-eyed suggestions, and Lizz Blaise and John Morrone did top-notch production and management. Rob Grom designed an incredibly eye-catching cover, and Steven Seighman worked on the interior design, which has been a dream come true. The whole St. Martin's team has been so wonderful to work with.

Befriending Elon Green has been one of my luckiest breaks in my whole career (and I've enjoyed a lot of luck). He introduced me to Laura and has so often given me crucial insight into the publishing industry. And without his assistance in archival access to old newspapers, I wouldn't have found the stories of William Tillman and Linda Joyce Glucoft. Not to mention his absolutely fantastic feedback on the book itself. I'm gonna get that landline someday!

My dear friend Maggie Bornholdt was instrumental at several points during the writing process. She alerted me to the Mario Markworth crimes soon after they happened in 2019, while I was writing my proposal. She was also my travel companion on a research trip in 2021, accompanying me to campuses, museums, RV parks, small-town Kentucky historical societies, relevant graveyards, and library archives. Alas, the

ACKNOWLEDGMENTS

chapter we primarily researched on this trip didn't end up in the manuscript, but even journeys down dead-end roads are rewarding with the right friend.

I also want to thank the other members of my writers' group, who have been here since the very beginning of this project. I started the group after a brainstorming session with Julia Gaughan, whose insight, wisdom, and compassion are such a great boon to not just me but our whole community. My nonfiction comrade Danny Caine inspires me with his diligent and lovely writing and his commitment to embodying his values. My neighbor Chance Dibben is the exact kind of cheerleader every book and every literary circle needs. Courtney Shipley often came to me with axe resources I would not have otherwise found. A few other local friends I'd like to mention: Althea Schnake, Will Averill, John Kaleugher, Richard Noggle, Allison Puderbaugh, Abby Olcese, Erin Schramm, Lyndsey Varella, Hailey Handy, Dan Hoyt, Nick Spacek, Brock Wilbur, and Laura Lorson, the booksellers at the Raven, my cousins Nancy and Cooper Sims-West, Lori Lange and the other members of my neighborhood book club, the whole city of Lawrence for that matter.

The friendships I made at Hollins University have lasted decades and sustained so much of my development as a writer. Miranda Dennis has been my sounding board for many of my questions from sentence structure to major editing choices. Rachel Emery, Carly Hays, and Moira Glace have listened to me rave about breakthroughs and vent about obstacles in the group chat for the duration. Special thanks to Rachel for styling and Carly for taking my author's photo in beautiful Margaritaville. A few other Hollins people I'd like to mention who offered help or just enthusiasm about the project: Carmen Sambuco and her husband Marshall Odell, Martha Sadler, Tif Robinette, Jessica Castigliego, Ashley Anderson, Elena Samel, Jennifer Elizabeth, Rhiannon Bly, and Emily Compton.

My boss at the University of Kansas, Jennifer Laverentz, has been such a boon to this project. Not only has she been flexible during crunch periods of the writing process, she invited both me and Jason to her home so that her husband Kale could give me a practical tutorial in chopping down a tree. My day job has been another piece of luck in more than one way. Access to the KU library system made the deep research that this book required not just possible but convenient. Not just the access to journals and web archives, but the fact that they brought pretty much any book I needed or wanted to my office often less than twenty-four hours after my request, and let me keep them often for years—well, all I can say is Rock Chalk Jayhawk!

I am deeply appreciative of all the scholars and experts who gave me their time and insight throughout the process. I've tried to quote most of them in the book, but many illuminating conversations didn't make it through the editing process. So now is the time to mention them by name: Molly Zahn, John Younger, Rabbi Zalman Tiechtel, Robbie Fox, Angelique Corthals, Rolf Quam, Noemi Sala, Garry Shaw, Tom Hulit, Josh Roberson, Sarah Chapman, Keren Wang, Yi Zhao, David Stuttard, Leszek Gardela, Oren Falk, Christine

Hernandez, Elizabeth Ashman Rowe, and Brett McLeod. I'd also like to thank the folks at the University of Kentucky Library Archives, the Livingston County Historical and Genealogical Society, the Metal Museum in Memphis, and tour guides at Monticello and Taliesin for their insight and guidance. Thanks also to Shaggy 2 Dope and the team at Psychopathic Records for granting permission to use their lyrics in my epigraph—it wouldn't have been right to have a book with so many hatchets and no ICP whatsoever.

The librarians at the Lawrence Public Library obtained and patiently waited for me to return dozens of interlibrary loans, grabbed and shelved my holds, helped me open up study rooms to conduct interviews, and a million other little things. Lorelei Corcoran, Vanessa Holden, Jennifer Elizabeth, Jonathan Earle, Ellen Morris, Lauren Nofi, and Valerie Hartman are just a few of the people who pointed a stranger in the right direction without any promise of reward. Maggie Serota also offered help early in the process. Sarah Weinman always offered sharp advice. Lyz Lenz initially commissioned an article for the Rumpus about why we use the phrase "axe murderer" and set off this project. Major thanks also to Mary Roach, Rachel Monroe, Daniel Stashower, and Gabrielle Moss for their generosity in writing blurbs.

My mom's delight in my successes and compassion for my failures—these are all you could hope for from a mother. And of course I wouldn't be in this position without my dad's support and insight (and his nepotism). Family and old friends I'd just like to mention because they have brightened my life: Isaac and Jessica James, Reuben James and

Jenny Stern, the Metzler family, Barry and Evan Graham, Cindy Danieley, Cindy DeJesus, Allison McCarthy, Aaron Layman, Eve Strillacci, Sara Lugar, Natacia Owens, the Echeverria family, Emily Kotay. I hope to see you all on the book tour!

SELECTED ADDITIONAL SOURCES

This book is slim—a hatchet, not a maul. But it took six years to write, and in the process I've consulted hundreds of books, journals, newspapers, and websites; conducted interviews, attended court dates, and visited crime scenes, but for the most part these chapters are based on repeated rounds of research. The sources I returned to again and again, the ones I had open while I was writing—these I strove to reference or quote in the text.

There were many quotes and digressions, however, that were shaved away even as they contributed to my overall understanding. This is especially true for the earlier chapters, when the subjects of inquiry were more ancient and outside of my initial frame of knowledge. By the later chapters I just wasn't going down quite as many dead-end roads, so there are a lot more additional sources for the early chapters.

"The Ancient Dominions of Maine: An Archaeology of Tools Historic Maritime I (1607–1676): The First Colonial Dominion." Davistown Museum, https://www.davistownmuseum.org/.

Arsuaga, J. L., et al. "Neandertal Roots: Cranial and Chronological Evidence from Sima de Los Huesos." *Science* 344, no. 6190 (2014): 1358–63, https://doi.org/10.1126/science.1253958.

"August 17—The End of Empson and Dudley." Uploaded by the Anne Boleyn Files and Tudor Society, YouTube, August 16, 2019, https://www.youtube.com/watch?v=2Up3Dpp2mIk.

Balcer, Jack Martin. "Herodotus, the 'Early State,' and Lydia." *Historia: Zeitschrift Für Alte Geschichte* 43, no. 2 (1994): 246–49, http://www.jstor.org/stable/4436327.

Barrett, James Harold. *Contact, Continuity, and Collapse: The Norse Colonization of the North Atlantic.* Brepols Publishers, 2003.

Bennett, Chris. "A Genealogical Chronology of the Seventeenth Dynasty." *Journal of the American Research Center in Egypt* 39, no. 123 (2002), https://doi.org/10.2307/40001152.

Berry, Daina Ramey. *The Price for Their Pound of Flesh: The Value of the Enslaved, from Womb to Grave, in the Building of a Nation.* Beacon Press, 2017.

Blumberg, Arnold. "Drawn and Quartered: The Law of Treason in Medieval England." *Medieval Warfare* 5, no. 2 (2015): 49–52.

Borneman, Walter R. *The French and Indian War: Deciding the Fate of North America.* Harper Perennial, 2007.

Borthwick, Mamah, and Alice T. Friedman. "Frank Lloyd Wright and Feminism: Mamah Borthwick's Letters to Ellen Key." *Journal of the Society of Architectural Historians* 61, no. 2 (2002): 140–51, https://doi.org/10.2307/991836.

Borthwick, Mark. *A Brave and Lovely Woman: Mamah Borthwick and Frank Lloyd Wright.* Madison: University of Wisconsin Press, 2023.

Carbonell, Eudald, and Marina Mosquera. "The Emergence of a Symbolic Behaviour: The Sepulchral Pit of Sima de Los Huesos,

Sierra de Atapuerca, Burgos, Spain." *Comptes Rendus Palevol* 5, nos. 1–2 (2006): 155–60, https://doi.org/10.1016/j.crpv.2005.11.010.

Cheung, Christina, et al. "Diets, Social Roles, and Geographical Origins of Sacrificial Victims at the Royal Cemetery at Yinxu, Shang China: New Evidence from Stable Carbon, Nitrogen, and Sulfur Isotope Analysis." *Journal of Anthropological Archaeology* 48 (December 2017): 28–45, https://doi.org/10.1016/j.jaa.2017.05.006.

Clark, Richard. "Being Hanged at Tyburn," n.d., http://www.capitalpunishmentuk.org/hangedt.html.

Ebrey, Patricia. "Bronzes from Fu Hao's Tomb." *UW Departments Web Server*, https://depts.washington.edu/chinaciv/archae/tfuhbron.htm.

Geserick, G., K. Krocker, and I. Wirth. "Über die Walcher'sche Hutkrempenregel—eine Literaturstudie [Walcher's Hat Brim Line Rule—a Literature Review]." *Arch Kriminol* 234, nos. 3–4 (2014): 73–90.

Girard, Philippe R. "French Atrocities during the Haitian War of Independence." *Journal of Genocide Research* 15, no. 2 (2013): 133–49.

Golden, Mark. 2008. "Helpers, Horses, and Heroes: Contests over Victory in Ancient Greece." *Greek Sport and Social Status*, University of Texas Press, 2008, ch. 1, pp. 1–39, https://doi.org/10.7560/718692-003.

Goldin, Paul R. "Some Shang Antecedents of Later Chinese Ideology and Culture." *Journal of the American Oriental Society* 137, no. 1 (December 16, 2021), https://doi.org/10.7817/jameroriesoci.137.1.0121.

Gray, Vivienne. "Herodotus' Literary and Historical Method: Arion's Story (1.23–24)," *American Journal of Philology* 122, no. 1 (2001): 11–28, https://doi.org/10.1353/ajp.2001.0008.

Grimal, Nicolas-Christophe, and Ian Shaw. *A History of Ancient Egypt.* Oxford: Blackwell, 1994.

Guerra, Maria F., and Sandrine Pagès-Camagna. "On the Way to the New Kingdom: Analytical Study of Queen Ahhotep's Gold Jewellery (17th Dynasty of Egypt)." *Journal of Cultural Heritage* 36 (March 2019): 143–52, https://doi.org/10.1016/j.culher.2018.09.004.

Hammond, N. G. L. "II. The Philaids and the Chersonese: I. The Three Bearers of the Name 'Miltiades.'" *Classical Quarterly* 6, nos. 3–4 (1956): 113–29, https://doi.org/10.1017/S0009838800020085.

Haywood, John. *Northmen: The Viking Saga, AD 793–1241*. St. Martin's Press, 2016.

Hendrickson, Paul. *Plagued by Fire: The Dreams and Furies of Frank Lloyd Wright*. Alfred A. Knopf, 2019.

Hightower-Langston, Donna. *The Native American World*. Wiley, 2002.

Horowitz, Mark R. "'Agree with the King': Henry VII, Edmund Dudley and the Strange Case of Thomas Sunnyff." *Historical Research: The Bulletin of the Institute of Historical Research* 79, no. 205 (2006): 325–66.

Impey, Edward, and Geoffrey Parnell. *The Tower of London: The Official Illustrated History*. Merrell Publishers in association with Historic Royal Palaces, 2000.

James, R. R. "Hanged, Drawn, and Quartered." *British Medical Journal* 2, no. 3474 (1927): 230.

Jennings, Francis. *Empire of Fortune: Crowns, Colonies, and Tribes in the Seven Years War in America*. W. W. Norton, 1988.

Keay, John. *China: A History*. Basic Books, 2011.

Keightley, David N. "At the Beginning: The Status of Women in Neolithic and Shang China." *NAN NÜ* 1, no. 1 (1999): 1–63.

Lee, Yun Kuen. "Building the Chronology of Early Chinese History." *Asian Perspectives* 41, no. 1 (2002): 15–42. http://www.jstor.org/stable/42928543.

Lengel, Edward G. *General George Washington: A Military Life*. New York: Random House, 2007.

Leon, Chrysanthi S. *Sex Fiends, Perverts, and Pedophiles: Understanding Sex Crime Policy in America.* New York University Press, 2011.

Lepore, Jill. *The Name of War: King Philip's War and the Origins of American Identity.* Vintage, 1999.

Lincoln, Victoria. *A Private Disgrace: Lizzie Borden by Daylight,* 1967.

Lindheim, Sara H. "Hercules Cross-Dressed, Hercules Undressed: Unmasking the Construction of the Propertian Amator in Elegy 4.9." *American Journal of Philology* 119, no. 1 (1998): 43–66, https://doi.org/10.1353/ajp.1998.0014.

Lodine-Chaffey, Jennifer Lillian. "Performing at the Block: Scripting Early Modern Executions." Ph.D. dissertation, University of Montana, 2013, https://scholarworks.umt.edu/etd/743.

Luraghi, Nino. *The Historian's Craft in the Age of Herodotus.* Oxford: Oxford University Press, 2001.

Lycett, Stephen J. "Understanding Ancient Hominin Dispersals Using Artefactual Data: A Phylogeographic Analysis of Acheulean Handaxes." *PLOS One* 4, no. 10 (2009): e7404. https://doi.org/10.1371/journal.pone.0007404.

Machin, Anna Jane. "Why Handaxes Just Aren't That Sexy: A Response to Kohn & Mithen (1999)." *Antiquity* 82, no. 317 (2008): 761–66, https://doi.org/10.1017/S0003598X00097362.

Marshall, Anthony J. "Symbols and Showmanship in Roman Public Life: The Fasces." *Phoenix* 38, no. 2 (1984): 120, https://doi.org/10.2307/1088896.

Martin, D. "Violence and Masculinity in Small-Scale Societies." *Current Anthropology*, 62, no. S23 (2021): S169–S181.

Milks, Annemieke. "What's in a Name? Defining Prehistoric Weaponry." *Sticks and Stones*, October 11, 2018, https://sticks-and-stones.blog/2018/10/11/whats-in-a-name-defining-prehistoric-weaponry/.

Muhlestein, Kerry. "Royal Executions: Evidence Bearing on the Subject of Sanctioned Killing in the Middle Kingdom." *Journal* of the Economic and Social History of the Orient 51, no. 2 (2008): 181–208, https://doi.org/10.1163/156852008x307429.

National Oceanic and Atmospheric Administration. "What Is a Rogue Wave?" National Ocean Service, 2019, https://oceanservice.noaa.gov/facts/roguewaves.html.

Park, Robert W. "Contact between the Norse Vikings and the Dorset Culture in Arctic Canada." *Antiquity* 82, no. 315 (2008): 189–98, https://doi.org/10.1017/s0003598x0009654x.

Parker, Victor. "Herodotus' Use of Aeschylus' Persae as a Source for the Battle of Salamis." *Symbolae Osloenses* 82, no. 1 (2007): 2–29.

Price, Neil, et al. "Viking Warrior Women? Reassessing Birka Chamber Grave Bj.581." *Antiquity* 93, no. 367 (2019): 181–98, https://doi.org/10.15184/aqy.2018.258.

Putt, Shelby S. "The Origins of Stone Tool Reduction and the Transition to Knapping: An Experimental Approach." *Journal of Archaeological Science: Reports* 2 (June 2015): 51–60, https://doi.org/10.1016/j.jasrep.2015.01.004.

Raffield, B. Playing Vikings. *Current Anthropology*, 60, no. 6 (2019), 813–35.

Rawson, Jessica. "Ancient Chinese Ritual Bronzes: The Evidence from Tombs and Hoards of the Shang (1500–1050 BC) and Western Zhou (1050–771 BC) Periods." *Antiquity* 67, no. 257 (December 1993): 805–23, https://doi.org/10.1017/s0003598x00063808.

Robin, Gerald D. "The Executioner: His Place in English Society." *British Journal of Sociology* 15, no. 3 (1964): 234–53, https://doi.org/10.2307/588468.

"Roofing hatchet?!" Uploaded by ThemRoofBoys, YouTube, 2022, https://www.youtube.com/shorts/qV0HeClwHfs.

Royer, Katherine. "The Body in Parts: Reading the Execution Ritual in Late Medieval England." *Historical Reflections / Réflexions Historiques* 29, no. 2 (2003): 319–39.

Russell, Gareth. "The Dudleys and the Royals They Served." *Tudor Life: The Tudor Society Magazine* 49 (2018).

Sabloff, Paula L. W. "How Pre-Modern State Rulers used Marriage to Reduce the Risk of Losing at War: A Comparison of Eight States." *Journal of Archaeological Method and Theory* 25, no. 2 (2108): 426–52, http://doi.org/10.1007/s10816-017-9342-2.

Sano, Katsuhiro, et al. "A 1.4-Million-Year-Old Bone Handaxe from Konso, Ethiopia, Shows Advanced Tool Technology in the Early Acheulean." *Proceedings of the National Academy of Sciences* 117, no. 31 (2020): 18393–400, https://doi.org/10.1073/pnas.2006370117.

Scheidel, Walter. "Slavery and Forced Labor in Early China and the Roman World." *SSRN Electronic Journal*, 2013, https://doi.org/10.2139/ssrn.2242322.

Schenawolf, Henry. "Tomahawks and Hatchets: Part 2 of 3—Trade Axes of America." *Revolutionary War Journal*, May 28, 2019, https://revolutionarywarjournal.com/tomahawks-and-hatches-part-2-of-3-trade-axes-of-america/.

Schulman, Alan R. *Ceremonial Execution and Public Rewards: Some Historical Scenes on New Kingdom Private Stelae*. Freiburg, Switzerland / Göttingen, Germany: Universitätsverlag / Vandenhoeck Ruprecht, 1988. Zurich Open Repository and Archive, University of Zurich, https://doi.org/10.5167/uzh-149722.

Shaw, Ian. *Ancient Egyptian Warfare*. Open Road Media, 2019.

Sidpura, Taneash. "Flies, Lions and Oyster Shells: Investigating Military Rewards in Ancient Egypt from the Predynastic Period to the New Kingdom (4000–1069 BCE)." Ph.D. dissertation, University of Manchester, 2022.

"Smell & Tell: The Smell of Mummies." *Ann Arbor District Library*, 2019, https://aadl.org/node/397485.

Stuttard, David. *A History of Ancient Greece in Fifty Lives*. National Geographic Books, 2014.

Surikov, I. "Nicknames among Greeks of the Archaic and Classical Periods: Preliminary Thoughts of a General Theoretical Nature." *Akropolis* 2, no. 1 (2018), 5–19.

Thorp, Robert L. *China in the Early Bronze Age Shang Civilization*.

University of Pennsylvania Press, 2013. Project Muse, https://muse.jhu.edu/chapter/866416.

Wang, Keren. "Atlas of Sacrifice: Three Studies of Ritual Sacrifice in Late-Capitalism." Ph.D. dissertation, University of Pennsylvania, 2018. ProQuest Dissertations and Theses Global, http://proquest.com.

Wang, Rong, Chang-sui Wang, and Ji-gen Tang. "A Jade Parrot from the Tomb of Fu Hao at Yinxu and Liao Sacrifices of the Shang Dynasty." *Antiquity* 92, no. 362 (April 2018): 368–82, https://doi.org/10.15184/aqy.2017.220.

"What Do Mummies Smell Like?" Uploaded by HMNS-Houston Museum of Natural Science, YouTube, May 14, 2019, https://www.youtube.com/watch?v=mYXdn2UEVnY.

Yates, Robin. "Slavery in Early China: A Socio-Cultural Approach." *Journal of East Asian Archaeology* 3, no. 1 (2001): 283–331, https://doi.org/10.1163/156852301100402723.

NOTES

INTRODUCTION

1. Associated Press. "Laborer Arrested in Colorado Axe-Murder." *Tampa Tribune*, September 22, 1911, p. 1, https://www.newspapers.com/image/326120666/.
2. Dostoyevsky, Fyodor. *Crime and Punishment*. Translated by Richard Pevear and Larissa Volokhonsky, New York: Alfred A. Knopf, 1993, p. 73.
3. Jackson, Shirley. *Life Among the Savages*. London: Penguin, 2020.
4. Friedman, Uri. "The North Korean Axe Murders That Almost Started a War." *The Atlantic*, June 10, 2018, https://www.theatlantic.com/international/archive/2018/06/axe-murder-north-korea-1976/562028.
5. Molyneux, Lizzie, and Wendy Molyneux. "Boyz 4 Now." *Bob's Burgers*, season 3, episode 21, Fox, April 28, 2013.
6. Barham, Lawrence. *From Hand to Handle: The First Industrial Revolution*. Oxford University Press, 2013, p. 3.
7. Orejel, Jorge L. "The "Axe/Comb" Glyph as ch' ak." Center for Maya Research, Washington, D.C., Research Reports on Ancient Maya Writing 31, 1990, pp. 1–8.
8. Berlin, Adele, and Marc Zvi Brettler, editors, Michael Fishbane, consulting editor. *The Jewish Study Bible: Jewish Publication Society Tanakh Translation*. Oxford: Oxford University Press, 2014, p. 389.
9. Ibid., p. 390.

NOTES

ONE: CRANIUM 17 AND THE PIT OF BONES

1. Corbey, R., A. Jagich, K. Vaesen, and M. Collard. "The Acheulean Handaxe: More Like a Bird's Song Than a Beatles' Tune?" *Evolutionary Anthropology* 25, no. 1 (2016): 6–19, https://doi.org/10.1002/evan.21467.
2. Kremer, C., et al. "Discrimination of Falls and Blows in Blunt Head Trauma: Systematic Study of the Hat Brim Line Rule in Relation to Skull Fractures." *Journal of Forensic Sciences*, 53, no. 3 (2008): 716–19.
3. Sala, N., et al. "Lethal Interpersonal Violence in the Middle Pleistocene." *PLOS One* 10, no. 5 (2015): e0126589, https://doi:10.1371/journal.
4. Interview with Rolf Quam.
5. Ibid.
6. Sala, et al.
7. Kohn, M., and S. Mithen. "Handaxes: Products of Sexual Selection?" *Antiquity* 73, no. 281 (1999): pp. 518–26.

TWO: THE SMITED KING

1. Morris, Ellen. "Daggers and Axes for the Queen: Considering Ahhotep's Weapons in Their Cultural Context." *The Treasure of the Egyptian Queen Ahhotep and International Relations at the Turn of the Middle Bronze Age, 1550 B.C.* Edited by G. Miniaci and P. Lacovara, London: Golden House Productions, 2022, pp. 165–86.
2. Wilkinson, Toby. *The Rise and Fall of Ancient Egypt*. New York: Random House, 2013.
3. HMNS-Houston Museum of Natural Science. "What Do Mummies Smell Like?" YouTube, 2019, https://www.youtube.com/watch?v=mYXdn2UEVnY.
4. Shaw, Garry J. "The Death of King Seqenenre Tao." *Journal of the American Research Center in Egypt* 45 (2009): 159–76, http://www.jstor.org/stable/25735452.
5. Saleem, Sahar N., and Zahi Hawass. "Computed Tomography Study of the Mummy of King Seqenenre Taa II: New Insights into His Violent Death." *Frontiers in Medicine* 8 (2021), doi:10.3389/fmed.2021.637527.
6. Feinman, Peter. Review of *The Quarrel Story: Egypt, the Hyksos, and Canaan. Conversations with the Biblical World* 35 (2015), https://www.academia.edu/37362108/The_Quarrel_Story_Egypt_the_Hyksos_and_Canaan.
7. Wilkinson, p. 185.
8. Ibid.

9. Ryholt, Kim, and Adam Bülow-Jacobsen. *The Political Situation in Egypt During the Second Intermediate Period, c. 1800–1550 B.C.* Copenhagen: Carsten Niebuhr Institute of Near Eastern Studies; Museum Tusculanum Press, 1997.
10. Wilkinson, p. 194
11. Morris, "Daggers and Axes for the Queen."

THREE: CASCADE OF BLOOD

1. Major, John S., and Constance A. Cook. *Ancient China: A History.* Abingdon, Oxon: Routledge, 2017, p. 77.
2. Fang-mei Chen. "The Bronze Weapons of the Late Shang Period." Ph.D. dissertation, University of London, 1997, p. 86.
3. Interview with Keren Wang, December 2021.
4. Majors and Cook, p. 79.
5. Linduff, Katheryn M., and Yan Sun. *Gender and Chinese Archaeology.* Walnut Creek, CA: AltaMira Press, 2004, p. 103.
6. Ibid., p. 21.

FOUR: IN TRUTH, AN ENEMY AND A MAN OF VIOLENCE

1. Purvis, Andrea, translator. *The Landmark Herodotus: The Histories.* Edited by Robert B. Strassler, Anchor Books, 2009.
2. Ibid.
3. Ibid.
4. Stuttard, David. *Phoenix: A Father, a Son, and the Rise of Athens.* Cambridge: Harvard University Press, 2021.
5. Immerwahr, H. R. "Stesagoras II." *Transactions and Proceedings of the American Philological Association* 103 (1972): pp. 181–86, https://doi.org/10.2307/2935974.
6. Unruh, Daniel. "Loaves in a Cold Oven: Tyranny and Sterility in Herodotus' Histories." *Classical World* 114, no. 3 (2021): 281–308, https://doi.org/10.1353/clw.2021.0012.
7. Herodotus. *The Histories.* Translated by A. D. Godley, bk. 6, ch. 38, sec. 1, accessed July 6, 2024, http://www.perseus.tufts.edu/hopper/text?doc=urn:cts:greekLit:tlg0016.tlg001.perseus-eng1:6.38.
8. Ibid.
9. Purvis, p. 442.
10. Scott, Lionel. *Historical Commentary on Herodotus*, bk. 6, Leiden: Brill, 2005.

11. Allgood, Evan. 2021. "Herodotus' Other Lies." *The New Yorker.* https://www.newyorker.com/humor/daily-shouts/herodotuss-other-lies.
12. Munson, Rosaria Vignolo. Introduction. Herodotus, *Herodotus and the World*, vol. 2, Oxford: Oxford University Press, 2013.
13. Waites, Margaret C. "The Deities of the Sacred Axe." *American Journal of Archaeology* 27, no. 1 (1923): 25–56, https://doi.org/10.2307/497531.
14. Davies, Malcolm. "Aeschylus' Clytemnestra: Sword or Axe?" *Classical Quarterly* 37, no. 1 (1987): 65–75, http://www.jstor.org/stable/639344.
15. Sophocles. *The Electra*. Edited by Richard Jebb, Cambridge University Press, 1894, https://www.perseus.tufts.edu/hopper/text.jsp?doc=Soph.+-El.+93.

FIVE: FREYDIS, WOMAN OF THE FOREST

1. "Ottar." Viking Ship Museum. Accessed August 28, 2024. https://www.vikingeskibsmuseet.dk/en/visit-the-museum/exhibitions/previous-exhibitions/heart-and-soul-50-years-with-the-ships-of-the-vikings-2012–13/exhibition-text/ottar.
2. Ferguson, Robert. *The Vikings: A History.* New York: Penguin Books, 2010, p. 280.
3. Parker, Philip. *The Northmen's Fury: A History of the Viking World.* London: Vintage Books, 2015, p. 172.
4. Eiricksson, Leifur, translator. *Vinland Sagas.* Penguin Books, 2008.
5. Reeves, Arthur. *The Finding of Wineland the Good.* London: H. Frowde, 1890.
6. Reeves. Arthur et al. *The Norse Discovery of America.* Norrœna Society. 1907, p. 232
7. Ibid.
8. Interview with Dr. Elizabeth Ashman Rowe.
9. Ibid.
10. Reeves, 1907, p. 233.
11. Ibid.
12. Reeves, 1907, p. 234.
13. Ibid.
14. Ibid.
15. Reeves, 1907, p. 234.
16. Eiricksson, p. 20.
17. Ibid.
18. Ibid.

19. Ibid.
20. Eiricksson, p. 45.
21. Falk, Oren. *The Bare-Sarked Warrior: A Brief Cultural History of Battlefield Exposure.* Tempe, AZ: ACMRS, Arizona Center for Medieval and Renaissance Studies, 2015.
22. Ibid.

SIX: PIGMEN, GARGOYLES, BLUNDERING YOUTHS

1. Cawthorne, Nigel. *Public Executions.* Edison, NJ: Chartwell Books, 2006, p. 36.
2. Ibid.
3. Gunn, Steven. *Henry VII's New Men and the Making of Tudor England.* New York: Oxford University Press, 2023, p. 8.
4. Ibid., p. 34.
5. Ibid., p. 9.
6. Cawthorne, Nigel. *Public Executions.* London: Capella, 2006, p. 100.
7. Blumberg, Arnold. "Drawn and Quartered: The Law of Treason in Medieval England." *Medieval Warfare* 5, no. 2 (2015): 49–52, https://www.jstor.org/stable/48578437.
8. Friedman, Toba Malka. "At the Block All Hero He Appear'd: Noble Execution and Redemption in Tudor England." Ph.D. dissertation, UCLA, 2009.
9. Harris, Barbara J. *Edward Stafford, Third Duke of Buckingham, 1478–1521.* Stanford University Press, 1986, p. 161.
10. Ibid., p. 162.
11. Guy, John. "Thomas More and Tyranny." *Moreana* 49 (no. 189–90), no. 3–4 (2012): 157–88, p. 167.
12. Lodine-Chaffey, Jennifer Lillian. *A Weak Woman in a Strong Battle: Women and Public Execution in Early Modern England.* Tuscaloosa: University of Alabama Press, 2022.
13. "Inside the Tower of London: The Tudor Tower." Films on Demand, Films Media Group, 2018, fod.infobase.com/PortalPlaylists.aspx?wID=104680&xtid=199270.
14. Friedman, p. 12.
15. Sargent, Daniel. *Thomas More.* Sheed and Ward, 1935. Internet Archive, accessed 2024, https://archive.org/details/thomasmore0000unse_h6h4/page/n7/mode/2up.
16. Howell, Thomas Bayly, et al. *A Complete Collection of State Trials and Proceedings for High Treason and Other Crimes and Misdemeanors from*

the Earliest Period to the Year 1783. London: Longman, 1816. Google Books, https://www.google.com/books/edition/A_Complete_Collection_of_State_Trials_an/2vc8UQII-jsC?hl=en.
17. Ibid.
18. Lodine-Chaffey, p. 25.
19. Ibid., p. 22.
20. Friedman, p. 119.
21. Ibid., p. 121.
22. Merriman, Roger Bigelow. *Life and letters of Thomas Cromwell, vol. 1 of 2 life, letters to 1535*. Project Gutenberg, 2015.
23. Wilson, Derek. *In the Lion's Court: Power, Ambition, and Sudden Death in the Reign of Henry VIII*. London: Vintage Digital, 2014, p. 463.
24. Elton, G. R. *The Tudor Revolution in Government*. Cambridge University Press, 1969.
25. Lodine-Chaffey, p. 42.
26. Russell, Gareth. *Young and Damned and Fair: The Life and Tragedy of Catherine Howard at the Court of Henry VIII*. London: William Collins, 2018, p. 318.
27. Lodine-Chaffey, p. 29.
28. Pierce, Hazel. "The Life, Career and Political Significance of Margaret Pole, Countess of Salisbury 1473–1541." Ph.D. dissertation, Bangor University (United Kingdom), 1997, p. 6.
29. Ibid., p.134.
30. Ibid., p. 314.
31. Friedman, p. 139.
32. Pierce, p. 314.

SEVEN: YOU ARE NOT DEAD YET, MY FATHER

1. Coe, Alexis. *You Never Forget Your First: A Biography of George Washington*. New York: Penguin, 2021, p. xxix.
2. Weems, Mason Locke. *The Life of George with Curious Anecdotes, Equally Honourable to Himself, and Exemplary to His Young Countrymen*. Philadelphia: Printed by M. Carey et Son, 1819.
3. Cook, D. *The Ax Book: The Lore and Science of the Woodcutter*. Chambersburg, PA: Alan C. Hood, 1999.
4. Stevens, Scott Manning. "Tomahawk: Materiality and Depictions of the Haudenosaunee." *Early American Literature* 53, no. 2 (2018): 475–511, https://doi.org/10.1353/eal.2018.0046.
5. Ibid.

6. Ibid.
7. Bruchac, Margaret, and Kayla Holmes. "Investigating a Pipe Tomahawk." Penn Museum Blog, November 9, 2018, https://web.archive.org/web/20240302230211/https://www.penn.museum/blog/museum/investigating-a-pipe-tomahawk/.
8. Anderson, Fred. *Crucible of War: The Seven Years' War and the Fate of Empire in British North America, 1754–1766.* New York: Vintage Books, 2001, p. 13.
9. Misencik, Paul R. George *Washington and the Half-King Chief Tanacharison: An Alliance That Began the French and Indian War.* Jefferson, NC: McFarland, 2014.
10. Calloway, Colin Gordon. *The Indian World of George Washington: The First President, the First Americans, and the Birth of the Nation.* New York: Oxford University Press, 2018. p. 72
11. Ibid.
12. Anderson, p. 50.
13. Anderson, p. 52.
14. Miscencik.
15. Miscencik.
16. Anderson, p. 60.
17. Miscencik.
18. Anderson, p. 61.
19. Coe, p. 20.
20. Kern, Kevin F., and Gregory S. Wilson. *Ohio: A History of the Buckeye State.* Hoboken, NJ: Wiley, 2013.

EIGHT: I SUPPOSE YOU KNOW WHAT I AM DOING

1. Gray, Thomas R.; Turner, Nat; and Royster, Paul (Depositor), "The Confessions of Nat Turner (1831)" (1831). Electronic Texts in American Studies. 15. https://digitalcommons.unl.edu/etas/15.
2. Ibid.
3. McGinty, Brian. *The Rest I Will Kill: William Tillman and the Unforgettable Story of How a Free Black Man Refused to Become a Slave.* New York: Liveright, 2017, ch. 1.
4. McGinty, ch. 1.
5. McGinty, ch. 3.
6. *Douglass' Monthly* 4, no. 3 (August 1861), https://nyheritage.contentdm.oclc.org/digital/collection/p15109coll7/id/173/rec/1.
7. McGinty, ch. 3.

8. Ibid.
9. Ibid.
10. McGinty, ch. 4.
11. Ibid.
12. Ibid.
13. Ibid.
14. *Douglass' Monthly.*
15. McGinty, ch. 7.

NINE: FIVE AXES IN THE CELLAR, ONE AXE ON THE ROOF

1. McCarthy, Susan. "Art and Artifacts in the Crusade for Prohibition in Kansas, 1854–1920." *Kansas History: A Journal of the Central Plains*, 45, no. 1, 2022.
2. James, Bill. *Popular Crime: Reflections on the Celebration of Violence.* New York: Scribner, 2012.
3. Tucher, Andie. *Froth and Scum: Truth, Beauty, Goodness, and the Ax Murder in America's First Mass Medium.* Chapel Hill: University of North Carolina Press, 1994.
4. "Brita Nelson: How She Butchered Her Husband Because She Disliked Him—Her Trial and Acquittal." *Chicago Daily Tribune*, June 5, 1876, 7.
5. "With Skulls And Hatchets." *The Sun* (New York City), June 14, 1893, p. 3, newspapers.com.
6. "That Hoodoo Hatchet." *Evening World* (New York City), June 12, 1893, p. 4, newspapers.com.
7. "Will Explode a Bomb." *Berkshire Eagle* (Pittsfield, MA), June 12, 1893, p. 1.
8. Faye, Musselman. "Timeline: The Day of the Borden Murders." *Fall River Herald News*, August 4, 2016, https://www.heraldnews.com/story/news/2016/08/04/timeline-day-borden-murders/27268980007/.
9. Miller, Sarah. *The Borden Murders: Lizzie Borden and the Trial of the Century.* New York: Schwartz and Wade Books, 2016, p. 193.
10. "Trial of Lizzie Borden, Volume 1." Lizzie Andrew Borden Virtual Museum and Library, accessed July 8, 2024, https://lizzieandrewborden.com/wp-content/uploads/2011/12/TrialBorden1.pdf.
11. Miller, p. 91.
12. Miller, p. 212.
13. Edholm, Steven. "Splitting Billets from Logs, Legit Tips & Tricks." YouTube, June 1, 2016, https://www.youtube.com/watch?v=eYqc0yPuSqY.
14. Miller, p. 131.

15. Ibid.
16. Showalter, Elaine, and English Showalter. "Victorian Women and Menstruation." *Victorian Studies* 14, no. 1 (1970): 83–89, http://www.jstor.org/stable/3826408.
17. Miller, p. 129.
18. Miller, p. 66.
19. Robertson, Cara. *The Trial of Lizzie Borden: A True Story*. New York: Simon & Schuster, 2019.
20. Miller, p. 66.
21. Miller, p. 194.
22. Robertson, p. 217.
23. Noe, Denise. "What Happened to the Weapon?" *The Hatchet: A Journal of Lizzie Borden and Victorian Studies*, July 5, 2018, https://lizzieandrewborden.com/HatchetOnline/what-happened-to-the-weapon.html.

TEN: A FULLER MEASURE OF LIFE AND TRUTH, AT ANY COST

1. Wright, Frank Lloyd. "The Art and Craft of the Machine." *Brush and Pencil* 8, no. 2 (1901): 77–90. https://doi.org/10.2307/25505640.
2. Drennan, William R. *Death in a Prairie House : Frank Lloyd Wright and the Taliesin Murders*. Madison: University of Wisconsin Press, 2008. p. 39.
3. Nissen, Anne. "From the Cheney House to Taliesin: Frank Lloyd Wright and Feminist Mamah Borthwick." Ph.D. dissertation, 1988, p. 14.
4. Nissen, p. 35.
5. Drennan, p. 60.
6. Drennan, p. 57.
7. Ibid.
8. Ibid., p. 123.
9. Ibid., p. 53.
10. McCrea, Ron. *Building Taliesin: Frank Lloyd Wright's Home of Love and Loss*. Madison: Wisconsin Historical Society Press, 2014.
11. Drennan, p. 61.
12. Ibid., p. 90.
13. Ibid., p. 143
14. Ibid., p. 139
15. Ibid., p. 112.
16. Ibid., p. 106.
17. Ibid., p. 102.

18. Ibid., p. 92.
19. Sebesta, Judith A. "Spectacular Failure: Frank Lloyd Wright's Midway Gardens and Chicago Entertainment." *Theatre Journal* (Washington, D.C.) 53, no. 2 (2001): 291–309.

ELEVEN: "WHOEVER COMES OVER, I GIVE ANYBODY CANDY"

1. Chandler, Raymond. *The Raymond Chandler Papers: Selected Letters and Non-Fiction, 1909–1959*. Edited by Tom Hiney, New York: Grove Press, 2002, p. 55.
2. "Stroble Trial." *Los Angeles Times,* January 5, 1950, p. 23.
3. "Ax-Wielder Kills L.A. Girl, 6; Manhunt Seeks Aged Suspect." *Los Angeles Mirror* 2, no. 31, November 15, 1949.
4. "Missing LA Girl, 6, Slain by Kidnaper." *Los Angeles Evening Citizen News*, November 15, 1949, p. 1.
5. "Child Murderer Seen Near Border." *Daily News* (Los Angeles), November 16, 1949.
6. "Psychiatrist Says Fear Behind Killing." *Daily News*, November 18, 1949, p. 48.
7. "Confession Details Story of Slaying." *Los Angeles Times*, November 22, 1949, p. 2.
8. "Daughter Relates Stroble's Habits." *Los Angeles Times*, November 17, 1949, p. 2.
9. "Fiend Murders 6-Year-Old LA Girl with Ax." *Daily News* (Los Angeles), November 15, 1949, p. 2.
10. "Confession Details Story of Slaying." *Los Angeles Times,* November 22, 1949, p. 2.
11. Ibid.
12. Ibid.
13. "Confession of Stroble Bared." *Los Angeles Mirror*, November 21, 1949, p. 5.
14. "Confession Details Story of Slaying." *Los Angeles Times*, November 22, 1949, p. 2.
15. "Fiend Murders 6-Year-Old LA Girl with Ax." *Daily News* (Los Angeles), November 15, 1949, p. 2.
16. "Mother Keeps Vigil." *Los Angeles Mirror,* November 15, 1949, p. 12.
17. "Seek Sex Killer." *Los Angeles Mirror*, November 15, 1949, p. 14.
18. Bruschke, Jon, and William Earl Loges. *Free Press vs. Fair Trials: Examining Publicity's Role in Trial Outcomes*. London: Routledge, 2005.
19. Jenkins, Philip. *Moral Panic: Changing Concepts of the Child Molester in Modern America*. New Haven: Yale University Press, 1998.

20. "L.A. Axe Murderer to Face Arraignment on Friday." *Vallejo Times-Herald*, November 22, 1949, p. 10.
21. "Psychiatrist Says Fear Behind Killing." *Daily News*, November 18, 1949, p. 48.
22. Ibid.
23. Davis, Sid. "Dangerous Stranger." Internet Archive. Accessed August 28, 2024. https://archive.org/details/dangerous_stranger.

TWELVE: CANDY

1. McLeod, Brett. *American Axe: Celebrating the Tool That Shaped a Continent.* North Adams, MA: Storey Publishing, 2020.
2. Bloom, John, and Jim Atkinson. *Evidence of Love: A True Story of Passion and Death in the Suburbs.* New York: Open Road Integrated Media, 2018, ch. 9.
3. Ibid.
4. Ibid.
5. Ibid.
6. Ibid., ch. 14.
7. Ibid.
8. Ibid.
9. Ibid., ch. 24.
10. Ibid.
11. Ibid.
12. Ibid.
13. Ibid, ch. 25.
14. Weiss, Jeffery. "Some in Wylie Don't Know of 1980 Ax Slaying; Others Can't Forget." *Dallas Morning News*, June 11, 2010.
15. Swanson, Doug. "20 Years After the Ax Slaying of Betty Gore, Victim's Daughters Still Have Questions." *Dallas Morning News*, June 11, 2000.
16. Bloom and Atkinson, ch. 2.
17. Swanson.
18. Ibid.

EPILOGUE

1. Stephen King, interviewed by Nathaniel Rich. "Stephen King, The Art of Fiction No. 189." *Paris Review*, Fall 2006, https://www.theparisreview.org/interviews/5653/the-art-of-fiction-no-189-stephen-king.

2. Serota, Maggie. "How 'So I Married an Axe Murderer' Bombed but Became a Classic Anyway." *SPIN,* July 6, 2024. https://www.spin.com/2018/09/so-i-married-an-axe-murderer-cast-writers-interviews/.
3. Court proceedings, 1916-CR, Jackson County Criminal Court, June 27, 2022.
4. Ibid.
5. Ibid.
6. Honeycutt, Sherae. "'He Didn't Deserve to Die': Homeless Man Killed in Double Stabbing Remembered as Kind, Loving." Fox 4, Kansas City, August 8, 2019, https://fox4kc.com/news/he-didnt-deserve-to-die-homeless-man-killed-in-double-stabbing-remembered-as-kind-loving/.
7. Rice, Glenn. "KC Man Killed in Ax Attack Was a Free-Spirited 'Hobo.'" *Kansas City Star,* August 8, 2019, https://www.kansascity.com/news/local/crime/article234078192.html.
8. Ibid.
9. Ibid.
10. Police report, 1916-CR, Jackson County Court, August 8, 2019.
11. Ibid.
12. Ibid.
13. Court proceedings.
14. Ibid.
15. Ibid.
16. Ibid.
17. Dostoyevsky, *Crime and Punishment,* p. 73.

ABOUT THE AUTHOR

Carly Hays

RACHEL McCARTHY JAMES was born and raised in Kansas, the daughter of baseball's Bill James and artist Susan McCarthy. She graduated from Hollins University in Roanoke, Virginia, where she studied writing and politics. Her first nonfiction book, *The Man from the Train*, was written in collaboration with her father. She lives with her husband, Jason, and pets in Lawrence, Kansas.